Marine Biotoxins and Seafood Poisoning

Marine Biotoxins and Seafood Poisoning

Special Issue Editors

Pedro Reis Costa
Jorge Diogène
Antonio Marques

MDPI • Basel • Beijing • Wuhan • Barcelona • Belgrade

MDPI

Special Issue Editors
Pedro Reis Costa
IPMA
Portugal

Jorge Diogène
IRTA
Spain

Antonio Marques
IPMA
Portugal

Editorial Office
MDPI
St. Alban-Anlage 66
4052 Basel, Switzerland

This is a reprint of articles from the Special Issue published online in the open access journal *Toxins* (ISSN 2072-6651) from 2018 to 2019 (available at: https://www.mdpi.com/journal/toxins/special issues/Biotoxin Seafood)

For citation purposes, cite each article independently as indicated on the article page online and as indicated below:

LastName, A.A.; LastName, B.B.; LastName, C.C. Article Title. *Journal Name* **Year**, *Article Number*, Page Range.

ISBN 978-3-03921-818-9 (Pbk)
ISBN 978-3-03921-819-6 (PDF)

Contents

About the Special Issue Editors

Pedro Reis Costa received his PhD degree in Marine Biology and Ecology from the University of Lisbon (Portugal, 2006), and continued as a Postdoctoral Fellow in NOAA Fisheries Science Center (Seattle, USA) in 2007–2009. He has been a Researcher at the Portuguese Institute of the Sea and Atmosphere (IPMA) since his appointment in 2009. He is mainly interested in the study of harmful algal blooms (HAB) and the transfer of marine toxins in the food web, including their fate and toxicological potential in marine/estuarine environments.

Jorge Diogène graduated in Biology at the Universitat de Barcelona (Spain, 1985) and holds an MSc in Marine Biology (University of South Alabama, USA, 1988) and PhD in Toxicology from the Université Paris 7 (France, 1993). He is Head of both the Unitat de Seguiment del Medi and the Marine and Continental Waters program at IRTA-Sant Carles de la Ràpita. His research interests are oriented towards the evaluation of toxin production of microalgae, toxin determination in shellfish, and the development of cell-based assays used to identify and quantify the toxicological potential of toxins. His responsibilities focus on the monitoring program for the quality of water in shellfish harvesting areas, as well as in the field of seafood quality analysis and food safety evaluation.

Antonio Marques received his PhD degree in Applied Biological Sciences from the University of Ghent (Belgium, 2005) and continued as a Researcher at the Portuguese Institute of the Sea and Atmosphere (IPMA) in 2006–2015. He has been a Senior Researcher at IPMA since his appointment in 2015. His main interests are in seafood quality and safety (chemical contaminants), innovative solutions for the seafood production and processing industries, seafood preservation methodologies, and climate change.

Editorial

Marine Biotoxins and Seafood Poisoning

Pedro Reis Costa [1], António Marques [1] and Jorge Diogène [2,*]

[1] IPMA-Portuguese Institute for the Sea and Atmosphere, Av. Alfredo Magalhães Ramalho 6, 1495-165 Lisbon, Portugal; prcosta@ipma.pt (P.R.C.); amarques@ipma.pt (A.M.)
[2] IRTA Institute of Agrifood Research and Technology, IRTA, Ctra. Villafranco, 5, 43540 Sant Carles de la Ràpita, Spain
*   Correspondence: Jorge.Diogene@irta.cat

Received: 9 September 2019; Accepted: 17 September 2019; Published: 24 September 2019

Prevalence of marine biotoxins in seafood has been associated with increasing frequency, intensity, and duration of harmful algal blooms, and an increase of the geographical and temporal distribution of harmful algae. New and emerging biotoxins have been recurrently detected in regions where they were previously absent, raising challenges to the economic sustainability of seafood production in coastal areas and to consumer health safety. The economic burden to seafood producers caused by the closure of production areas and a possible feeling of insecurity from consumers urges researchers to improve available knowledge on toxin dynamics in marine organisms and the environment. Epidemiological studies are scarce and risk characterization is needed, particularly for emerging toxins. It is critical to enhance collaborative multi- and trans-disciplinary actions to introduce eco-innovative sustainable strategies to improve shellfish and fish safety. Strengthening industrial competitiveness is achievable by developing fast and reliable methods for marine biotoxin detection, and by implementing mitigation strategies. Innovative toxicological approaches for seafood safety evaluation are also required. The seven articles of this special issue address such research needs and are organized into three groups: (i) toxin dynamics and effects in marine organisms, (ii) development of detection methods for marine toxins, and (iii) toxin exposure and risks associated with the consumption of contaminated seafood.

Within the first group of articles, Alvarez et al. [1] reported an extreme event in southern Chile of a bloom of *Alexandrium catenella* that caused mass mortality of several marine invertebrate species, and resulted in accumulation of high levels (exceeding 100 times the regulatory limit for human consumption) of paralytic shellfish poisoning (PSP) toxins in clams. This study highlights the need for assessing toxin dynamics in shellfish under controlled conditions to better understand and foresee the impacts of harmful algal blooms. Andres et al. [2] fed green-lipped mussel (*Perna viridis*) with the toxic dinoflagellate *Alexandrium minutum* under controlled laboratory conditions to assess the dynamics of PSP toxin levels during accumulation and elimination phases. Barbosa et al. [3] investigated the interaction of ocean warming with fish (*Sparus aurata*) exposure to PSP toxins through contaminated mussels to assess physiological responses and changes in toxin accumulation. Tetrodotoxin (TTX), which has a mode of action comparable to PSP toxins, was characterized in the greater blue-ringed octopus *Hapalochlaena lunulata* from Okinawa, Japan [4].

Regarding the development and optimization of methods for toxin detection, Chen and colleagues [5] optimized clean-up procedures based on immunoaffinity column purification before mass spectrometry detection, providing an improved way to detect the amnesic shellfish poisoning toxin domoic acid (DA), in an array of matrices. Lefebvre et al. [6] describes a DA-specific antibody in the human serum and report DA-chronic exposure to certain groups of shellfish consumers. Finally, Hayashi et al. [7] investigated the combined effect of okadaic acid and mycotoxins that are considered emerging toxins in the marine environment, in human intestinal cell lines.

Acknowledgments: Gratitude is due to all contributing authors and reviewers.

Conflicts of Interest: The authors declare no conflict of interest.

References

1. Álvarez, G.; Díaz, P.; Godoy, M.; Araya, M.; Ganuza, I.; Pino, R.; Álvarez, F.; Rengel, J.; Hernández, C.; Uribe, E.; et al. Paralytic Shellfish Toxins in Surf Clams *Mesodesma donacium* during a Large Bloom of *Alexandrium catenella* Dinoflagellates Associated to an Intense Shellfish Mass Mortality. *Toxins* **2019**, *11*, 188. [CrossRef] [PubMed]
2. Andres, J.; Yñiguez, A.; Maister, J.; Turner, A.; Olano, D.; Mendoza, J.; Salvador-Reyes, L.; Azanza, R. Paralytic Shellfish Toxin Uptake, Assimilation, Depuration, and Transformation in the Southeast Asian Green-Lipped Mussel (*Perna viridis*). *Toxins* **2019**, *11*, 468. [CrossRef] [PubMed]
3. Barbosa, V.; Santos, M.; Anacleto, P.; Maulvault, A.L.; Pousão-Ferreira, P.; Costa, P.R.; Marques, A. Paralytic Shellfish Toxins and Ocean Warming: Bioaccumulation and Ecotoxicological Responses in Juvenile Gilthead Seabream (*Sparus aurata*). *Toxins* **2019**, *11*, 408. [CrossRef] [PubMed]
4. Asakawa, M.; Matsumoto, T.; Umezaki, K.; Kaneko, K.; Yu, X.; Gomez-Delan, G.; Tomano, S.; Noguchi, T.; Ohtsuka, S. Toxicity and Toxin Composition of the Greater Blue-Ringed Octopus *Hapalochlaena lunulata* from Ishigaki Island, Okinawa Prefecture, Japan. *Toxins* **2019**, *11*, 245. [CrossRef] [PubMed]
5. Chen, S.; Zhang, X.; Yan, Z.; Hu, Y.; Lu, Y. Development and Application of Immunoaffinity Column Purification and Ultrahigh Performance Liquid Chromatography-Tandem Mass Spectrometry for Determination of Domoic Acid in Shellfish. *Toxins* **2019**, *11*, 83. [CrossRef] [PubMed]
6. Lefebvre, K.; Yakes, B.; Frame, E.; Kendrick, P.; Shum, S.; Isoherranen, N.; Ferriss, B.; Robertson, A.; Hendrix, A.; Marcinek, D.; et al. Discovery of a Potential Human Serum Biomarker for Chronic Seafood Toxin Exposure Using an SPR Biosensor. *Toxins* **2019**, *11*, 293. [CrossRef] [PubMed]
7. Hayashi, A.; José Dorantes-Aranda, J.; Bowman, J.P.; Hallegraeff, G. Combined Cytotoxicity of the Phycotoxin Okadaic Acid and Mycotoxins on Intestinal and Neuroblastoma Human Cell Models. *Toxins* **2018**, *10*, 526. [CrossRef] [PubMed]

toxins

MDPI

Article

Paralytic Shellfish Toxin Uptake, Assimilation, Depuration, and Transformation in the Southeast Asian Green-Lipped Mussel (*Perna viridis*)

John Kristoffer Andres [1],*, **Aletta T. Yñiguez [1]**, **Jennifer Mary Maister [1]**, **Andrew D. Turner [2]**,
Dave Eldon B. Olano [1], **Jenelyn Mendoza [1]**, **Lilibeth Salvador-Reyes [1]** and **Rhodora V. Azanza [1]**

[1] The Marine Science Institute, University of the Philippines Diliman, Quezon City 1101, Philippines
[2] Food Safety Group, Centre for Environment, Fisheries and Aquaculture Science, Barrack Road, Weymouth, Dorset DT4 8UB, UK
* Correspondence: jkristofferandres@gmail.com; Tel.: +63-2-636-9355

Received: 5 July 2019; Accepted: 6 August 2019; Published: 9 August 2019

Abstract: Bivalve molluscs represent an important food source within the Philippines, but the health of seafood consumers is compromised through the accumulation of harmful algal toxins in edible shellfish tissues. In order to assess the dynamics of toxin risk in shellfish, this study investigated the uptake, depuration, assimilation, and analogue changes of paralytic shellfish toxins in *Perna viridis*. Tank experiments were conducted where mussels were fed with the toxic dinoflagellate *Alexandrium minutum*. Water and shellfish were sampled over a six day period to determine toxin concentrations in the shellfish meat and water, as well as algal cell densities. The maximum summed toxin concentration determined was 367 µg STX eq./100 g shellfish tissue, more than six times higher than the regulatory action limit in the Philippines. Several uptake and depuration cycles were observed during the study, with the first observed within the first 24 h coinciding with high algal cell densities. Toxin burdens were assessed within different parts of the shellfish tissue, with the highest levels quantified in the mantle during the first 18 h period but shifting towards the gut thereafter. A comparison of toxin profile data evidenced the conversion of GTX1,4 in the source algae to the less potent GTX2,3 in the shellfish tissue. Overall, the study illustrated the temporal variability in *Perna viridis* toxin concentrations during a modelled algal bloom event, and the accumulation of toxin from the water even after toxic algae were removed.

Keywords: saxitoxin; harmful algal blooms; biotransformation; uptake; depuration; assimilation; shellfish; *Perna viridis*; *Alexandrium*

Key Contribution: The study shows the rapid uptake of toxins in *P. viridis* as well as its ability to convert from higher potency to lower potency saxitoxin analogues.

---

1. Introduction

Paralytic shellfish poisoning (PSP) is caused by the consumption of shellfish such as bivalve molluscs contaminated with paralytic shellfish toxins (PST), a family of compounds related to saxitoxin (STX) which are produced naturally by several species of dinoflagellates. Uptake and depuration of toxins within the flesh of the molluscs varies greatly from one shellfish species to another [1–5], with toxin retention lasting from days to months, depending on the species [5]. Differences in the accumulation of PST between different bivalve species has been reported during a bloom of *Pyrodinium bahamense* var. *compressum* (PBC) in Masinloc Bay, Philippines [6]. Seven species of bivalves were tested for toxicity where six of them became toxic. The species *Spondylus squamosus* obtained the highest toxicity during the peak of the bloom and several cycles of uptake and depuration of toxin were observed within a year. *S. squamosus* was also able to uptake toxins even in the absence of the causative phytoplankton, with toxicity eventually increasing following another algal bloom. The same behaviour was observed in *Atrina vexellum* but at lower magnitudes. As for the green-lipped mussel, *Perna viridis*, the pattern for toxicity seemed to follow the cycle for PBC wherein during the onset of the bloom, toxicity was seen to increase in the shellfish tissue, then decreasing when the bloom of PBC declined.

Toxins are taken inside the shellfish system through filter-feeding activity but are not distributed evenly across different parts of the tissue [2]. Depending on the time of toxin exposure, the burden or amount of overall toxicity per part shifts from one organ to another. Generally, the viscera (organs in the abdominal cavity including the digestive gland), show the highest toxicity [1,2,7]. A five-compartment model has been developed for *Perna viridis* fed with *Alexandrium tamarense* [7]. The mussel was dissected into five parts, namely the hepatopancreas, viscera, gill, adductor muscle, and foot. The highest toxin levels were determined in the hepatopancreas (47–74% of toxicity), followed by the viscera (8–41%), gill (2–18%), adductor muscle (1–13%), and foot (0.4–5%). Interestingly, the organ group with the lowest contribution to the total shellfish body mass had the highest level of toxicity. The high relative concentrations of toxins in the hepatopancreas relate to the fact that this organ is responsible for toxin removal from the shellfish system. The distribution of PST in *Perna viridis* and the scallops *Chlamys nobilis* fed with *Alexandrium tamarense* has also been assessed [8]. The mussel was dissected into the digestive gland and other parts while the scallop was dissected into adductor muscle and other tissues. Results highlighted that in mussels, the digestive gland contained the highest toxin burden. As for the scallops, toxin burden was higher in the other tissues compared to the adductor muscle. Two phases of depuration were observed for both shellfish species. The first phase was characterised by fast depuration, which was presented as the gut evacuation of the unassimilated toxins. The second phase was slower, which was thought related to be depurating toxins becoming assimilated and incorporated into other tissues.

Transformation of saxitoxin analogues has been observed by monitoring the differences in toxin profiles detected in both the source phytoplankton and shellfish tissue [2]. Biotransformation is thought to usually occur during periods of detoxification or contamination [9]. Changes in toxin profiles within shellfish tissues may occur either through selective retention of specific toxin analogues, or through enzymatic transformation, indicating active toxin metabolism in shellfish [10]. PSTs undergo transformation from one form to another through different processes. Such processes include reduction, epimerisation, oxidation, and, desulfation, all of which may potentially result in changes to overall shellfish toxicity [3], following the conversion of toxins to other analogues of lower or higher potency. Conversion to less potent forms may accelerate detoxification through the flushing out of toxin from the shellfish system, while conversion to more potent forms will result in a slower detoxification rate [9]. Many studies have shown the ability of *Perna viridis* as well as other species such as *Chlamys nobilis*, to convert STX analogues to more potent forms [11]. Epimerisation has also been detected through the conversion of predominantly C2 in *A. tamarense* to C1 in both shellfish species [8]. In addition, new metabolites were found in the shellfish that were not present in the source phytoplankton.

Different methods are used to detect and quantify PST, including animal bioassays, chemical detection methods and immunoassays [12]. The mouse bioassay (MBA) has for many decades been the reference method used for toxin monitoring programmes in most countries and for many years has been the official control reference method in both the EU and the US. Chemical detection methods such as high-performance liquid chromatography with fluorescence detection (HPLC-FLD) enable the quantitation of individual toxins or epimeric pairs, facilitating the calculation of total sample toxicity using appropriate toxicity equivalence factors (TEFs). To date two separate HPLC-FLD methods have been developed and validated through collaborative study, both using sample derivatization. The first, post-column oxidation HPLC-FLD was developed in Canada and has been accepted for official control testing by the ISSC for use in the US and Canada [13]. The second, known as the Lawrence method, utilizes pre-column oxidation with HPLC-FLD and starting 1st January 2019 became the official reference method for PSP analysis within the EU [14]. More recently mass spectrometry-based methods have been developed and validated [15] enabling the rapid quantitation of a wider range of analogues.

HABs occur globally throughout all the major continents, with the South China Sea known to be one of the world's hotspots for HAB occurrences [16]. However, there are limited studies in the region involving the dynamics and kinetics of HAB toxins, particularly PST in bivalves. The major toxin producers for this region are *Alexandrium sp.* including species originally assigned as *Gonyaulux sp.* [8]. These species produce primarily N-sulfocarbamoyl PST and C toxins, with C2 as the dominant toxin type. There is minimal information, however, on the toxin dynamics of *Perna viridis* as it interacts with toxic *Alexandrium* species. Since *P. viridis* is one of the main contributors to aquaculture production in the Philippines, and also serves as a cheap source of food for the local population, it is important to determine the risk of toxicity from this species both from a health and socio-economic perspective. Consequently, this study sought to assess the pattern of uptake and depuration of toxins within the mussels, the distribution patterns of toxins within the shellfish tissue and the presence and implications of toxin transformation. These results, obtained through laboratory feeding experiments, would then be used to assess the overall risks from PST in mussels from the Philippines.

2. Results

2.1. Overall Toxicity, Uptake, and Depuration

The total PST concentrations were used to estimate total toxicity in each of the shellfish samples analysed. Mussels sampled from the negative control tanks (Tank C containing shellfish only) were found to contain low levels of GTX1,4; and GTX2,3 analogues due to their prior exposure in the field. Therefore, the average toxicity values obtained from the negative control tanks were subtracted from the values of toxicity from the phytoplankton + shellfish set-up for each sampling period to give an accurate estimation of toxin uptake. The regulatory maximum permitted limit (MPL) for PST in the Philippines is 60 µg saxitoxin equivalents per 100 g shellfish meat (600 µg STX eq./kg). Total shellfish toxin concentrations determined here were generally lower than the MPL within the first six hours of the feeding study. From the 12th hour sampling point until the end of the experiment, total PST exceeded the MPL with the highest total toxin levels occurring in mussels sampled after 96 h, containing 367 (±166.17) µg STX eq./100 g (Figure 1a and Supplementary Material Table S1).

The uptake and depuration rates for total PST in mussels following exposure to toxic *A. minutum* cycled three times (Figure 1b). The first cycle of uptake and depuration occurred within the first 24 h of the experiment. The second and third cycles of uptake were then repeated during the 48th hour and 96th hour of sampling, respectively, while depuration occurred at the 72nd and 120th hour. Interestingly, the uptake rates before depuration occurred had similar values (from 8–11 µg STX eq./h), implying a threshold uptake rate for shellfish.

The mussels within the experimental tanks can potentially accumulate toxins from two sources: (1) the *A. minutum* cells; and (2) any toxins dissolved in the water. The *A. minutum* cell densities in tank A initially increased but began to decrease at the 12th hour. By the 72nd hour onwards, there were no cells present since the water in the tanks was changed to fresh seawater, free from toxic algae (Figure 2a). The cell densities in tank B, the phytoplankton control, showed a similar trend at the start. However, the cells remained beyond the 72nd hour since water here was not changed. In the water column of the tanks, overall toxicity exhibited some increases and decreases (Figure 2b and Supplementary Material Table S2). Tank B showed a peak in toxin levels during the 72nd hour, while for the phytoplankton + shellfish tank, a peak occurred at the 48th hour and a smaller peak was also observed during the 12th hour. Variability in the toxin concentrations was high and no significant difference was observed in the overall toxin concentrations in water samples between the phytoplankton only tanks (tank B) and those with phytoplankton + shellfish (tank A). The average toxin value recorded for water in tank B was 11.1 µg STX eq./100 mL (range: 1–26 µg STX eq./100 mL), while for the water in tank A, the average toxicity was higher at 23.5 µg STX eq./100 mL (range: 3–67 µg STX eq./100 mL).

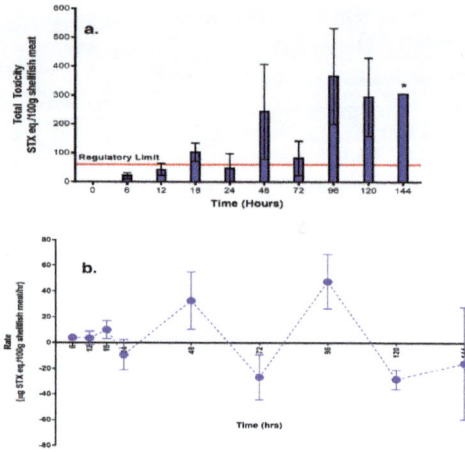

Figure 1. Paralytic shellfish toxins (PST) quantity and patterns of uptake and depuration for the shellfish in tank A (phytoplankton + shellfish set-up) (**a**) total toxicity; (**b**) rate. (* no standard error since only one tank was sampled due to shellfish mortality).

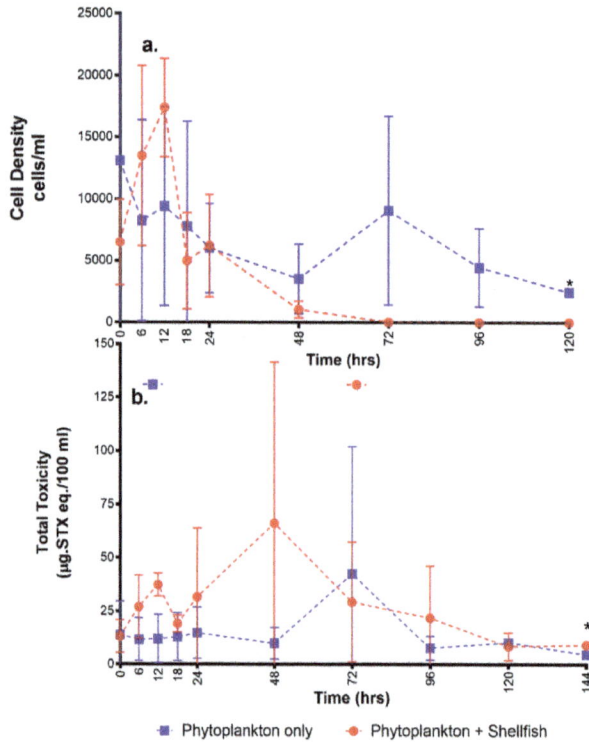

Figure 2. Parameters measured in the water: (**a**) cell density; (**b**) total toxicity. (* no standard error since only one tank was sampled due to shellfish mortality).

2.2. Toxin Compartmentalisation

The toxin distributions quantified within the shellfish tissues varied temporally and between parts (Figure 3 and Supplementary Material Figure S1). Six hours after feeding, the mantle and the gut contained similar toxin burdens at 52% and 47%, respectively. The part of the mussels containing the lowest toxin concentrations was the foot and adductor muscle (0.7%). The same toxin burden pattern was seen from the 12th hour to the 18th hour. From the 24th hour of the experiment to the 144th hour, toxin burden shifted more towards the gut (59–91%), followed by the mantle (8–40%), and again, the least burden was determined in the muscle (0.12–1.05%). Significant differences in the overall toxicity values between the gut and muscle and mantle and muscle were observed (Kruskal–Wallis, *p*-value < 0.05). No significant difference in the toxin concentration was found between gut and mantle

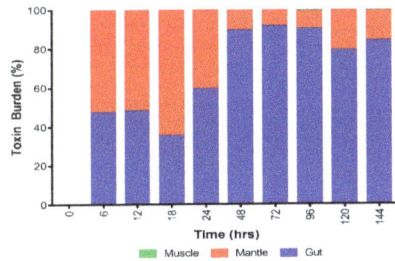

Figure 3. The distribution of toxin at different parts of the shellfish through time.

2.3. Toxin Types and Their Distributions

The toxin profile of *A. minutum* from the phytoplankton only set-up was dominated by GTX1,4; with traces of GTX2,3 (Figure 4 and Supplementary Material Table S3). In the tanks containing phytoplankton and shellfish, the toxin profiles were reversed with GTX2,3 dominating; and with lower relative proportions of GTX1,4.

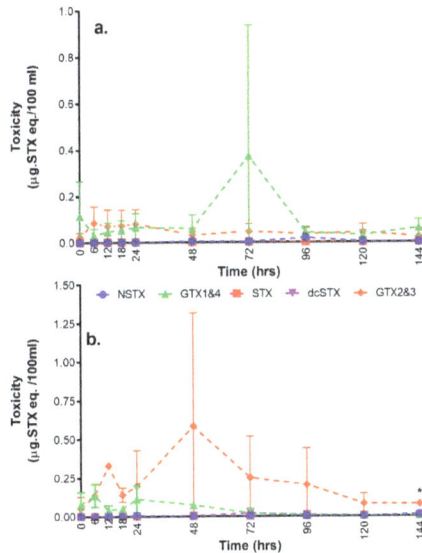

Figure 4. The patterns of saxitoxin analogues in water through time at (**a**) phytoplankton only set-up and (**b**) phytoplankton + shellfish. (* no standard error since only one tank was sampled due to shellfish mortality).

The PST analogue concentrations determined in the shellfish meat are summarised in Table 1. Representative chromatograms can be found in Figure S2 of the Supplementary Material GTX1,4; and GTX2,3 showed the highest toxin concentration values, whilst other analogues were present at either low concentrations or were non-detectable. Thus, GTX1,4; and GTX2,3 were used to more closely investigate the toxicity patterns in the shellfish meat.

Table 1. Overall toxin concentrations in *P. viridis* for each PST analogue through time (mean ± S.E. in µg STX eq./100 g).

Time (h)	NeoSTX	STX	dcSTX	GTX 1,4	GTX 2,3
0	n.d.	n.d.	n.d.	0 ± 0.14	0 ± 0.04
6	n.d.	n.d.	n.d.	12.15 ± 3.05	9.71 ± 7.82
12	n.d.	0.60 ± 0.68	n.d.	22.14 ± 3.90	20.55 ± 17.57
18	n.d.	15.82 ± 14.45	0.10 ± 0.02	50.98 ± 31.95	36.30 ± 13.30
24	n.d.	0.78 ± 0.13	0.02 ± 0.03	26.37 ± 23.53	20.07 ± 28.44
48	n.d.	0.91 ± 0.66	0.01 ± 0.02	77.13 ± 70.38	164.9 ± 120.4
72	n.d.	0.50 ± 0.15	0.02 ± 0.02	8.05 ± 15.54	73.76 ± 44.09
96	n.d.	4.81 ± 1.81	n.d.	59.86 ± 54.36	302.1 ± 155.7
120	n.d.	2.82 ± 1.74	n.d.	183.6 ± 146.6	108.4 ± 62.92
144	n.d.	4.27 ± 2.47	n.d.	105.9 ± 61.12	195.5 ± 112.8

n.d.: not detectable

Looking more closely at the uptake and depuration rates for GTX1,4; and GTX2,3; three uptake cycles were also apparent (Figure 5). From the start of the experiment until the 96th hour, uptake and depuration for both toxins occurred synchronously. However, during the 120th hour, a further increase in GTX1,4 was observed whilst conversely there was a depuration loss of GTX2,3. A reverse in uptake and depuration between two GTX analogues was also observed at the 144th hour when uptake was observed for GTX2,3; and depuration occurred for GTX1,4.

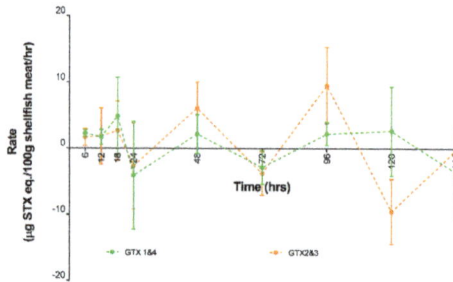

Figure 5. Uptake and depuration rate for GTX1,4; and GTX2,3.

The gut and muscle showed the same toxin analogue profile where GTX2,3 tended to have higher concentrations relative to GTX1,4. The highest average toxicity recorded for the gut was 271 µg STX eq./100 g shellfish meat, while the muscle had a much lower value of 2.8 µg STX eq./100 g. In the gut, GTX2,3 was relatively higher already at the 48th hour onwards, while toxicity in the muscle became more pronounced during the latter part of the study (Figure 6a,c). For the mantle, GTX1,4 had the highest concentration at the start of the experiment from 0–48 h. Toxin analogues shifted towards GTX2,3 from 72nd hour onwards (Figure 6b).

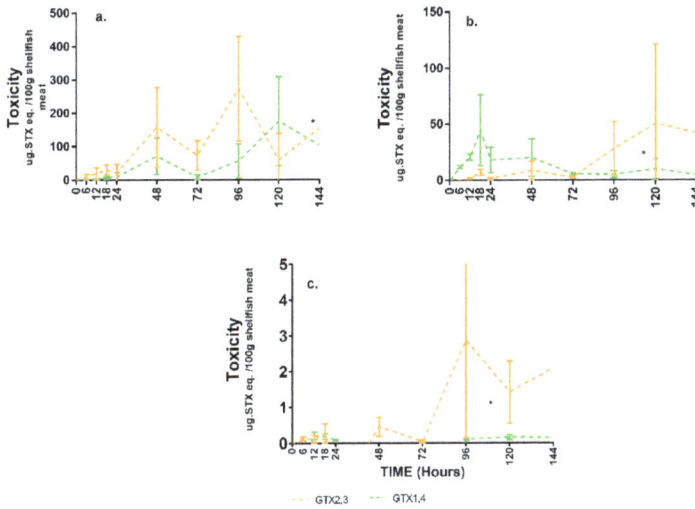

Figure 6. Concentration of GTX1,4 and GTX2,3 through time at (**a**) gut; (**b**) mantle; and (**c**) muscle. (*no standard error since only one tank was sampled due to shellfish mortality).

Qualitatively the difference in terms of overall water toxicity and PST analogues can be seen, though statistically there were no significant differences (average total toxicity and saxitoxin analogues between times; multiple *t*-test). This is likely due to the high variability between shellfish samples and/or the lower power of non-parametric statistics. This high inter-individual variability among shellfish has been observed frequently [3,7,8,17–19] and one potential explanation is the diverse physiological processes leading to sensitivity of shellfish to PST toxins [17].

3. Discussion

3.1. Uptake and Depuration Patterns

In a period less than 24 h, toxin uptake began, suggesting that this species can respond rapidly to the presence of the toxic phytoplankton. Bivalve feeding rhythms are thought to be based on food availability, with previous work demonstrating that there are cycles of clearance rate which are seen to occur within 24 h [11]. This cyclical pattern of uptake and depuration may represent the natural feeding behaviour of shellfish. Within the first 6 h of the feeding study, toxicity was generally lower than the national regulatory MPL. However, subsequent measurements starting from the 12th hour showed total mussel toxicity to rise above this limit. Based on the cycles of uptake and depuration, *P. viridis* is capable of depurating within a period of 24 hours. Moreover, repeated cycles of uptake and depuration were observed up to the 144th hour, the last sampling period. Due to the limitations in experimental design, the time to which the toxin will return below the MPL cannot be determined. From previous reports, however, toxin accumulation and elimination are relatively rapid in mussels, taking from days to weeks. In scallops and butter clams, elimination is known to be slower, sometimes over one year, due to toxin binding within the siphon tissue of animals [2,3,7]. One of the few studies that looked at short-term responses examined the uptake of Azaspiracids (AZAs) in scallops and mussels. The initial uptake was recorded from 24–48 h of the study, followed by depuration between 48–72 h. The cycle of uptake and depuration was recorded again during the next sampling periods with the same interval [20]. Moreover, the cycles for uptake and depuration show a similar pattern to the levels of total toxicity recorded. Interestingly, the results of our study showed that there appears to be a threshold value for the uptake rate (8–11 µg STX eq./hr) before depuration proceeds. This pattern can also be the result of the one-time feeding in the experiment, and could also represent conditions in the field wherein there is a pulse of high phytoplankton concentration which then declines as the bloom dies off. The highly variable toxin concentrations can pose a problem for HAB toxin monitoring programmes since the measured toxin would be dependent on the timing relative to the uptake and depuration cycle of the mussels. In addition, water containing toxic cells was removed on the 72nd hour in the tank A set-ups and were replaced with toxin-free UV-sterilized seawater. However, the shellfish were still able to accumulate toxins even after the removal of toxic cell source. Since the set-up was a closed-system, these toxins may have come from the dead mussels or depuration from other mussels, which could then have been taken up again by the remaining *P. viridis*.

3.2. Toxin Distribution within Shellfish

The toxins quantified in the shellfish taken from the toxic shellfish tank system were not equally distributed. This is important to determine given that the mussels are harvested and separated into specific parts prior to consumption [5]. For the first 18 h of the study, toxin burden in the mantle was the highest with 51–64% of the total toxin content. This was followed by the gut and lastly by the muscle group with 35–47% and 0.51–0.78% toxicity respectively. Changes in toxin profiles were observed after 24 h wherein the gut contained the highest toxin levels (59–91%). The mantle had 8–40% while the muscle group had 0.12–1.05% toxin burden recorded up until the last sampling period. The patterns observed here conform to the general observation that PSP toxins tend to accumulate the highest in the viscera [1,3–5] and that the distribution of toxins among tissues may shift depending on the time of exposure to toxins [4]. Despite the low contribution of viscera to the total soft tissue mass, it still has the highest toxicity. The viscera is the first organ exposed to the toxic cells and where toxins are absorbed, ingested, and metabolized [5]. In contrast, the foot and adductor muscle contribute to the majority of the soft tissue mass of the shellfish but was found to have the lowest toxicity. Depending on the time of exposure to toxic phytoplankton, the toxin levels present in each part of the molluscs changed, resulting in changes to toxin profiles between compartments. In other previous studies describing the toxification phases of *S. giganteus*, *M. mercenaria*, *S. solidissima*, *P. magellanicus*, and *Mya arenaria*, the viscera was reported to contain the highest toxin burden and shifted towards the siphon-gill tissue during detoxification phases. In *Mytilus edulis*, the viscera contained the highest toxin concentrations throughout the process [5]. The results of this study therefore conformed to the general observations that the gut had the highest toxin burden [1,3–5,8]. This may be attributed to the role of the gut to breakdown toxins from toxic algae. These observations therefore highlight the importance of appropriate mussel preparation when processing contaminated mussels for human consumption.

3.3. Toxin Transformation

Certain specific bivalve species are able to biotransform PST analogues following uptake of toxins from phytoplankton during filter-feeding, resulting in the generation of metabolites not found within the source algae [2,5,10,21,22]. Due to practical limitations in relation to access to commercial reference standards, only five saxitoxin analogues were assessed during this study. These analogues are considered to be the most potent PSTs. The dominant analogue detected in the water of the tanks containing only *A. minutum* was GTX1,4 with only traces of GTX2,3. Algal profiles were therefore similar to those of some previous studies [23], although toxin profiles in *A. minutum* are known to vary significantly throughout the world falling into a number of distinct clusters [24]. Conversely, however, GTX2,3 was found to be higher in the water sampled from the tanks containing shellfish and phytoplankton, as well as in the tissue from the shellfish in the same tanks. This suggests that biotransformation had occurred to convert GTX1,4 to GTX2,3. Analysis of samples was performed during the same analytical batch demonstrating these variations were not related to any aspect of method repeatability. Similarly, a shift in the toxin analogues detected in *Perna viridis* has been previously reported following feeding with *A. fundyense*. The source algae had a high ratio of GTX2/GTX3, but this ratio decreased in the shellfish samples indicating active epimerization within the shellfish tissue, prior to later transformation to STX through reduction [25]. Complex mechanisms involved in selective accumulation and chemical/enzymatic processes may be involved in the shellfish toxicity development [26]. These changes in analogues usually occur during periods of detoxification or contamination [9]. Different processes for toxin biotransformation include reduction, epimerization, oxidation, and desulfation [2–5,21,22]. The biotransformation potentially observed in this study resulted in the conversion to a less potent (GTX 2,3) from a more potent (GTX1,4) form. Transformation from GTX1,4 to GTX2,3 may be due to the epimerization process [27], and/or with the help of enzymes [28]. In addition, transformation is reported through reductive elimination with the decrease of N–OH group and increase in N–H group and elimination of sulfate group at the C11 position [29]. These transformations could be attributed to the shellfish itself or possibly bacterial action inside the shellfish [17] with the exact mechanisms in *P. viridis* requiring further investigation. Ideally any future work should incorporate a full suite of analytical standards into the detection method, with LC-MS/MS being utilised for a better understanding of the biotransformation processes within the shellfish tissues. Overall, the toxin profile data quantified in algae and mussels shows good evidence for transformation of toxin analogues, indicating the potential for species-specific transformation reactions similar to a number of other bivalve mollusc species.

From a more speculative ecological perspective, areas with high seawater residence time including many HAB-affected embayments in the Philippines, may encounter conditions where the toxin remains in the area due to the poor flushing of the water. Prolonged shellfish bans and toxicity in the shellfish have been observed in the Philippines even when the blooms of toxic cells have disappeared. These could be due to the ingestion of cysts [18,30] or illustrated here, through toxins that are present in the water column. This suggests that in our HAB monitoring programmes it will be advisable not only to measure HAB species cell densities, but also, at least initially, the toxin concentrations present in the water using technologies such as SPATT [31]. In addition, temporal and within-shellfish variabilities in toxin concentration can be further considered in designing HAB toxin monitoring programmes since the toxin levels measured could significantly change depending on the geographical and temporal nature of the sampling conducted.

4. Conclusions

Harmful algal blooms can plague certain areas in the Philippines for extended periods of time, resulting in a number of shellfish species accumulating toxins that present significant health and socio-economic risks. Here we show that in less than 24 h, *P. viridis* was able to uptake toxins when fed with toxic algae and consistently remained above the regulatory MPL in the Philippines from the 12th hour of the study onwards. Uptake and depuration rates cycled several times, which could represent a threshold value for the rate when shellfish starts to depurate and/or be related to the inconsistent supply of toxic *A. minutum* cells used for feeding. Even though toxic cells were removed, an increase in toxin concentrations in the shellfish were observed, which likely came from the toxin available in the water. This toxin could originate from dead mussels, faeces, and/or the depuration of toxin from the remaining toxin-containing mussels. In terms of the toxin distribution throughout the mussel tissue compartments, toxin profiles varied depending on the time of exposure to toxic cells. Initially, the mantle had the highest toxin burden, but later shifted towards the gut after 24 h. The toxin profiles of the source phytoplankton were found to be different from the profiles determined in both the shellfish tissue itself and the water sampled from the tanks containing shellfish fed with phytoplankton. GTX1,4 were dominant in the source algae with traces of GTX2,3, while the reverse was observed in both shellfish meat and water of the shellfish fed with phytoplankton setup. This indicates that biotransformation potentially occurred leading to a less toxic form.

Overall this study confirms the importance of both a water and shellfish monitoring programme for managing the risks from paralytic shellfish toxins in mussels from the Philippines. The current sampling design in toxin monitoring programmes of HAB affected areas should likely be re-assessed in order to help ensure food safety. Another key consideration is to test levels of toxins dissolved in seawater in addition to cell counts and shellfish toxicity. This may help shed light on observations of prolonged shellfish bans even if toxic cells were no longer detected. Similar studies are also needed for other shellfish species commonly consumed in the country.

5. Materials and Methods

5.1. Cell Cultures

Cultures of *Alexandrium minutum* were obtained from standing stock cultures (A.minBat). Cells were then sub-cultured to bigger flasks every three weeks until the volume reached six 80 L of culture of *A. minutum*. Culturing of cells to reach the final volume took approximately 18 weeks to be completed.

5.2. Acclimation

Perna viridis mussels were obtained from Bolinao, Pangasinan where HAB events occur regularly. Mussels were acclimatised for four days in a seawater aquarium with aerators, with temperature and salinity regulated to 27 °C and 32 ppt, respectively, whilst being fed daily with cultures of non-toxin producing *Isochrysis galbana*. Water was replaced every day to prevent accumulation of mucus that can cause mussel death. After 4 days, mussels were starved for 24 h prior to the grazing experiment commencing.

5.3. Grazing Experiment

Three different treatments were implemented in 80 L aquarium tanks: Tank type A contained mussels with *A. minutum* (phytoplankton + shellfish), tank type B contained mussels only (shellfish) and tank type C contained *A. minutum* only (phytoplankton) (Figure 7). Three replicate tanks were set up per treatment resulting in a total of nine 80 L tanks. Eighty mussels were placed into each of the first six tanks (tank set-ups A and B). In the first treatment (A) shellfish were fed with toxic *A. minutum* using concentrations of approximately 8000 cells/mL. Feeding was only done at the start of the experiment with the exposure to toxic cells for two days only. After the 48th hour sampling, toxic cells were removed and the tank water replaced with filtered seawater. The second treatment (B) served as a negative control and contained 80 individuals of shellfish per replicate without feeding of any toxic algae. The last treatment (C) served as a positive control containing *A. minutum* in concentrations of approximately 8000 cells/mL, without the presence of any shellfish. All tanks had UV-filtered seawater with constant aeration. Sampling was performed every six hours for 24 h and daily thereafter up to seven days. Five individual mussels from each shellfish-containing tank were sampled at each sampling period. These mussels were then divided into three parts for toxin analysis: gills + mantle + gonads, foot + adductor muscle; and, lastly the digestive gland or gut. Individuals were dissected over ice to minimise temperature effects on toxin changes. Shellfish meat samples were then stored in a freezer until required for toxin analysis. All parts of the five individuals were pooled together before toxin analysis.

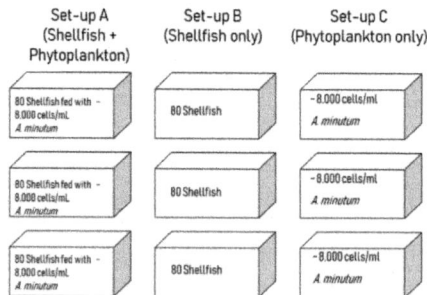

Figure 7. Schematic diagram of the experiment set-ups.

5.4. Toxin Analysis and Cell Counts

Toxin extraction of shellfish meat was conducted with the use of a ratio of 1:1 shellfish meat to 1 mL of 0.1 M hydrochloric acid [32]. Acid was added to tissue samples, vortex mixed then placed in a boiling water bath for five minutes. After boiling, the extract pH was adjusted to 3, centrifuged and supernatants filtered through a 0.2 µm syringe filter. Solid phase extraction (SPE) was used to clean-up the extracts as per Lawrence and Menard (1991) with modifications. A total of 250 µL of derivatizing agent (0.03 M HIO_4, 0.3 M NH_4HCO_3, 0.3 M Na_3PO_4) with a ratio of 1:1:1, and 50 µL from the shellfish samples were incubated for 3 min [32]. CH_3COOH (5.0 µL) was added prior to injection of a 50 µL aliquot of the reaction mix into the HPLC (LC-10A Shimadzu HPLC with RF-10AXL fluorescence detector). Separation used an Inertsil ODS-3V C18 (4.6 × 150 mm) column at 0.90 mL min^{-1} flow with binary gradient solvent system of 0.1 M ammonium formate (pH 7.0) for solvent A and HPLC-grade acetonitrile for solvent B. The elution gradient consisted of: 1% solvent B for 2 min, 1–5% solvent B for 3 min; 5% solvent B for 4 min; 5–6% solvent B for 1 min; 6–10% solvent B for 10 min.

Detection of PST analogues was conducted through comparison of oxidation product chromatographic peaks against those generated from working calibration standards prepared from Certified Reference Materials (CRMs) of STX, neo-STX, dcSTX, GTX1,4; and GTX2,3. CRMs were obtained from the Institute of Biotoxin Metrology, National Research Council, Canada. Chromatography and FLD was conducted according to AOAC 2005.06 [33].

Water (100 mL) was collected from the centre of each tank during every sampling period and used for toxin analysis. An additional 3 mL was collected near the surface, mid-depth, and bottom part of the tanks and used for phytoplankton counting to determine algal cell density through time. Cell counts were used to determine the clearance rate over time of mussel individuals.

Uptake and depuration rates were calculated by assessing the total toxicity accumulated over time by adding total toxin from each compartment within the shellfish tissues. The following equation was used for this computation

$$\text{Uptake Rate/Depuration Rate} = (\text{TTCN} - \text{TTCN-1})/h \tag{1}$$

where: TTCN = Total toxicity Concentration at time N.

Assimilation of toxin was measured by assessing the total toxicity present in the gut and other shellfish parts over time.

5.5. Data Analysis

The resulting data did not conform to assumptions of homogeneity and normality despite various transformations, thus non-parametric tests were used for statistical analyses. A multiple *t*-test was used to determine significant differences at each time point for overall toxicity of water, cell densities, in the phytoplankton only and shellfish fed with phytoplankton setups, and GTX values at different parts and in the water. A Kruskal–Wallis test was used to determine significant difference in overall toxicity across shellfish parts.

This paper is derived from the corresponding author's master's thesis [34].

Supplementary Materials: The following are available online at http://www.mdpi.com/2072-6651/11/8/468/s1, Table S1: Total toxicity in shellfish samples. Table S2: Toxicity of water in phytoplankton+shellfish set-up and phytoplankton only. Table S3: Toxin analogues of *A. minutum* and *P. viridis*. Figure S1: Image showing the parts of *Perna viridis* that were used for the study (Digestive Gland or Gut; Gills and Mantle; Adductor Muscle and Foot). Figure S2: Representative chromatograms from the HPLC for the Muscle (**A**), Gut (**B** and **C**), and Mantle (**D**). Standards for the toxins are overlaid on the diagrams and their peaks labeled.

Author Contributions: Conceptualization, J.K.A., A.T.Y., R.V.A., L.S.-R., J.M.M., D.E.B.O., A.D.T.; methodology, J.K.A., J.M.M.; resources, J.K.A., J.M.M., J.M., A.T.Y., R.V.A., L.S.-R.; HPLC analysis, J.K.A., J.M.M., J.M., D.E.B.O.; data curation, J.K.A., J.M.M., J.M., D.E.B.O.; writing—original draft preparation, J.K.A., A.T.Y.; writing-review and editing, A.D.T., A.T.Y.; funding acquisition, A.T.Y., J.K.A.; supervision, A.T.Y.; project administration, J.K.A., A.T.Y.

Funding: The research was funded by the University of the Philippines Diliman In-House Project and the following Department of Science and Technology-Philippine Council for Agriculture, Aquatic, and Natural Resources Research and Development (DOST-PCAARRD) projects: Operational Predictive System for Philippine Harmful Algal Blooms and Hazard Detection and Mitigation Tools for Algal Blooms in a Changing Marine Environment.

Acknowledgments: The researcher would like to acknowledge UP Marine Science Institute for housing the experiment, BiOME Laboratory and Red Tide Laboratory for the resources.

Conflicts of Interest: The authors declare no conflict of interest. The funders had no role in the design of the study; in the collection, analyses, or interpretation of data; in the writing of the manuscript, or in the decision to publish the results.

References

1. Mak, K.C.; Li, A.M.; Hsieh, D.P.; Wong, P.S.; Lam, M.H.; Wu, R.S.; Richardson, B.J.; Lam, P.K. Paralytic shellfish toxins in green-lipped mussels, *Perna viridis*, in Hong Kong. *Mar. Pollut. Bull.* **2003**, *46*, 258–268. [CrossRef]
2. Ding, L.; Qiu, J.; Li, A. Proposed biotransformation pathways of new metabolites of paralytic shellfish toxins based in field and experimental mussel samples. *J. Agric. Food Chem.* **2017**, *65*, 5494–5502. [CrossRef] [PubMed]
3. Kwong, R.W.; Wang, W.X.; Lam, P.K.; Peter, K.N. The uptake, distribution and elimination of paralytic shellfish toxins in mussels and fish exposed to toxic dinoflagellates. *Aquat. Toxicol.* **2006**, *80*, 82–91. [CrossRef] [PubMed]
4. Yu, K.N.; Choi, M.C.; Shen, X.; Wu, R.; Wang, W.; Lam, P. Modeling of depuration of paralytic shellfish toxins in *Chlamys nobilis* and *Perna viridis*. *Mar. Pollut. Bull.* **2005**, *50*, 463–484. [CrossRef] [PubMed]
5. Bricelj, V.M.; Shumway, S. Paralytic shellfish toxins in bivalve molluscs: Occurrence, transfer kinetics, and biotransformation. *Rev. Fish. Sci.* **1998**, *6*, 315–383. [CrossRef]
6. Montojo, U.; Sakamoto, S.; Cayme, M.; Gatdula, N.; Furio, E.; Relox, J.; Sato, S.; Fukuyo, Y.; Kodama, M. Remarkable difference in accumulation of paralytic shellfish toxins among bivalve toxins exposed to *Pyrodinium bahamense var. compressum* bloom in Masinloc Bay, Philippines. *Toxicon* **2006**, *48*, 85–92. [CrossRef]
7. Li, A.M.; Yu, P.K.; Hsieh, D.P.; Wang, W.-X.; Wu, R.S.; Lam, P.K. Uptake and depuration of paralytic shellfish toxins in green-lipped mussel, *Perna viridis*: A dynamic model. *Environ. Toxicol. Chem.* **2005**, *24*, 129–135. [CrossRef]
8. Choi, M.-C.; Hsieh, D.P.H.; Lam, P.K.S.; Wang, W.-X.; Lam, K.S.P.; Wang, W. Field depuration and biotransformation of paralytic shellfish toxins in scallop *Chlamys nobilis* and green-lipped mussel *Perna viridis*. *Mar. Biol.* **2003**, *143*, 927–934. [CrossRef]
9. Guéguen, M.; Baron, R.; Bardouil, M.; Truquet, P.; Haberkorn, H.; Lassus, P.; Barillé, L.; Amzil, Z. Modelling of paralytic shellfish toxin biotransformation in the course of *Crassostrea gigas* detoxification kinetics. *Ecol. Model.* **2011**, *222*, 3394–3402. [CrossRef]
10. Contreras, A.; Marsden, I.; Munro, M. Effects of short-term exposure to paralytic shellfish toxins on clearance rates and toxin uptake in five species of New Zealand bivalve. *Mar. Freshw. Res.* **2011**, *63*, 166–174. [CrossRef]
11. Wong, W.H.; Cheung, S.G. Feeding behaviour of the green mussel, *Perna viridis* (L.): Responses to variation in seston quantity and quality. *J. Exp. Mar. Biol. Ecol.* **1999**, *236*, 191–207. [CrossRef]
12. Costa, P.R.; Baugh, K.A.; Wright, B.; Ralonde, R.; Nance, S.L.; Tatarenkova, N.; Etheridge, S.M.; Lefebvre, K.A. Comparative determination of paralytic shellfish toxins (PSTs) using five different toxin detection methods in shellfish species collected in the Aleutian Islands, Alaska. *Toxicon* **2009**, *54*, 313–320. [CrossRef] [PubMed]
13. Van De Riet, J.; Gibbs, R.S.; Muggah, P.M.; Rourke, W.A.; MacNeil, J.D.; Quilliam, M.A. Liquid Chromatography Post-Column Oxidation (PCOX) Method for the Determination of Paralytic Shellfish Toxins in Mussels, Clams, Oysters, and Scallops: Collaborative Study. *J. AOAC Int.* **2011**, *94*, 1154–1176. [PubMed]
14. Turner, A.D.; Hatfield, R.G.; Maskrey, B.H.; Algoet, M.; Lawrence, J.F. Evaluation of the New European Union Reference Method for Paralytic Shellfish Toxins in Shellfish: A Review of Twelve Years Regulatory Monitoring Using Pre-Column Oxidation LC-FLD. *Trends Anal. Chem.* **2019**, *113*, 124–139. [CrossRef]
15. Turner, A.D.; McNabb, P.S.; Harwood, D.T.; Selwood, A.I.; Boundy, M.J. Single laboratory validation of a multitoxin LC-hydrophilic interaction LC-MS/MS method for quantitation of Paralytic Shellfish Toxins in bivalve shellfish. *J. AOAC Int.* **2015**, *98*, 609–621. [CrossRef] [PubMed]

16. Li, S.; Wang, W.; Hsieh, D. Feeding and absorption of the toxic dinoflagellate *Alexandrium tamarense* by two marine bivalves from the South. China Sea. *Mar. Biol.* **2001**, *139*, 617–624. [CrossRef]

17. Smith, E.A.; Grant, F.; Ferguson, C.M.J.; Gallacher, S. Biotransformations of paralytic shellfish toxins by bacteria isolated from bivalve molluscs. *Appl. Environ. Microbiol.* **2001**, *67*, 2345–2353. [CrossRef] [PubMed]

18. Persson, A.; Smith, B.C.; Wikfors, G.H.; Quilliam, M. Grazing on toxic *Alexandrium fundyense* resting cysts and vegetative cells by the eastern oyster (Crassostrea virginica). *Harmful Algae* **2006**, *5*, 678–684. [CrossRef]

19. Sombrito, E.; Honrado, M.; de Vera, A.; Tabbada, R.; Rañada, M.L.; Relox, J., Jr.; Tangonan, M.D. Use of *Perna viridis* as a Bioindicator of Paralytic Shellfish Toxins at Low *Pyrodinium bahamense* var *compressum* Density using a Radioreceptor Assay. *Environ. Bioindic.* **2007**, *2*, 264–272. [CrossRef]

20. Ji, Y.; Qiu, J.; Xie, T.; McCarron, P.; Li, A. Accumulation and transformation of azaspiracids in scallops (*Chlamys nobilis*) and mussels (*Mytilus galloprovincialis*) fed with *Azadinium popurum*, and response of antioxidant enzymes. *Toxicon* **2018**, *143*, 20–28. [CrossRef]

21. Asakawa, M.; Beppu, R.; Ito, K.; Tsubota, M.; Takayama, H.; Miyazawa, K. Accumulation of paralytic shellfish poison (PSP)and biotransformation of its components in oysters *Crassostrea gigas* fed with the toxic dinoflagellate *Alexandrium tamarense*. *Shokuhin Eiseigaku Zasshi* **2005**, *47*, 28–32. [CrossRef]

22. Botelho, M.J.; Vale, C.; Mota, A.; Simoes-Goncalves, M. Depuration kinetics of paralytic shellfish toxins in *Mytilus galloprovincialis* exposed to *Gymnodinium catenatum*: Laboratory and field experiments. *J. Environ. Monit.* **2010**, *12*, 2269–2275. [CrossRef] [PubMed]

23. Homan, N.; Hallegraeff, G.; Van Ruth, R.; Van Ginkel, P.; McNabb, P.; Kiermeier, A.; Deveney, M.; McLeod, C. *Uptake, Distribution and Depuration of Paralytic Shellfish Toxins in Australian Greenlip Abalone, Haliotis laevigata*; Australian Seafood Cooperative Research Centre: Deakin, Australia, 2010; pp. 1–26.

24. Lewis, A.M.; Coates, L.N.; Turner, A.D.; Percy, L.; Lewis, J. A review of the global distribution of *Alexandrium minutum* (Dinophyceae) and comments on ecology and associated paralytic shellfish toxin profiles, with a focus on Northern Europe. *J. Phycol.* **2018**, *54*, 581–598. [CrossRef] [PubMed]

25. Yu, K.N.P.; Kwong, R.W.; Wang, W.-X.; Lam, P.K.; Lam, K.S.P. Biokinetics of paralytic shellfish toxins in the green-lipped mussel *Perna viridis*. *Mar. Pollut. Bull.* **2007**, *54*, 1031–1071. [CrossRef] [PubMed]

26. Manfrin, C.; de Moro, G.; Torboli, V.; Venier, P.; Pallavacini, A.; Gerdol, M. Physiological and molecular responses of bivalves to toxic dinoflagellates. *Invertebr. Surviv. J.* **2012**, *9*, 184–199.

27. Ichimi, K.; Suzuki, T.; Yamasaki, M. Non-selective retention of PSP toxins by the mussel *Mytilus galloprovincialis* fed with the toxic dinoflagellate *Alexandrium tamarense*. *Toxicon Off. J. Int. Soc. Toxinol.* **2002**, *39*, 1917–1921. [CrossRef]

28. Wiese, M.; D'Agostino, P.M.; Mihali, T.K.; Moffitt, M.C.; Neilan, B.A. Neurotoxic Alkaloids: Saxitoxin and Its Analogs. *Mar. Drugs* **2010**, *8*, 2185–2211. [CrossRef] [PubMed]

29. Jaime, E.; Gerdts, G.; Luckas, B. In vitro transformation of PSP toxins by different shellfish tissues. *Harmful Algae* **2007**, *6*, 308–316. [CrossRef]

30. Yñiguez, A.T.; Maister, J.; Villanoy, C.L.; Deauna, J.D.; Peñaflor, E.; Almo, A.; David, L.T.; Benico, G.A.; Hibay, E.; Mora, I.; et al. Insights into the dynamics of harmful algal blooms in a tropical estuary through an integrated hydrodynamic-*Pyrodinium*-shellfish model. *Harmful Algae* **2018**, *80*, 1–14. [CrossRef]

31. Kudela, R. Passive sampling for freshwater and marine algal toxins. *Compr. Anal. Chem.* **2017**, *78*, 379–409.

32. Lawrence, J.; Menard, C. Determination of marine toxins by liquid chromatography. *J. Anal. Chem.* **1991**, *339*, 494–498. [CrossRef]

33. Horwitz, W.; Latimer, G. *Official Methods of Analysis*, 18th ed.; Method 2005.06; AOAC INTERNATIONAL: Gaithersburg, MD, USA, 2005; Chapter 49.

34. Andres, J.K. Saxitoxin Uptake, Assimilation, Depuration, and Analogue Changes in Perna viridis (Linnaeus). Unpublished Master's Thesis, University of the Philippines Diliman, Quezon City, Philippines, 2019.

toxins

MDPI

Article

Paralytic Shellfish Toxins and Ocean Warming: Bioaccumulation and Ecotoxicological Responses in Juvenile Gilthead Seabream (*Sparus aurata*)

Vera Barbosa [1,2], **Marta Santos** [1,2], **Patrícia Anacleto** [1,2,3], **Ana Luísa Maulvault** [1,2,3], **Pedro Pousão-Ferreira** [4], **Pedro Reis Costa** [1,5] **and António Marques** [1,2,*]

[1]   IPMA—Portuguese Institute for the Sea and Atmosphere, I.P., Av. Doutor Alfredo Magalhães Ramalho, n° 6, 1495-165 Algés, Portugal
[2]   CIIMAR—Interdisciplinary Centre of Marine and Environmental Research, University of Porto, Terminal de Cruzeiros do Porto de Leixões, Avenida General Norton de Matos S/N, 4450-208 Matosinhos, Portugal
[3]   MARE—Marine and Environmental Sciences Centre, Guia Marine Laboratory, Faculty of Sciences, University of Lisbon (FCUL), Av. Nossa Senhora do Cabo, 939, 2750-374 Cascais, Portugal
[4]   IPMA—Portuguese Institute for the Ocean and Atmosphere, EPPO-Aquaculture Research Station, 8700-305 Olhão, Portugal
[5]   CCMAR—Centre of Marine Sciences, University of Algarve, Campus of Gambelas, 8005-139 Faro, Portugal
*   Correspondence: amarques@ipma.pt; Tel.: +351-21-3027000

Received: 4 June 2019; Accepted: 9 July 2019; Published: 13 July 2019

Abstract: Warmer seawater temperatures are expected to increase harmful algal blooms (HABs) occurrence, intensity, and distribution. Yet, the potential interactions between abiotic stressors and HABs are still poorly understood from ecological and seafood safety perspectives. The present study aimed to investigate, for the first time, the bioaccumulation/depuration mechanisms and ecotoxicological responses of juvenile gilthead seabream (*Sparus aurata*) exposed to paralytic shellfish toxins (PST) under different temperatures (18, 21, 24 °C). PST were detected in fish at the peak of the exposure period (day five, 0.22 µg g^{-1} N-sulfocarbamoylGonyautoxin-1-2 (C1 and C2), 0.08 µg g^{-1} Decarbamoylsaxitoxin (dcSTX) and 0.18 µg g^{-1} Gonyautoxin-5 (B1)), being rapidly eliminated (within the first 24 h of depuration), regardless of exposure temperature. Increased temperatures led to significantly higher PST contamination (275 µg STX eq. kg^{-1}). During the trial, fish antioxidant enzyme activities (superoxide dismutase, SOD; catalase, CAT; glutathione S-transferase, GST) in both muscle and viscera were affected by temperature, whereas a significant induction of heat shock proteins (HSP70), Ubiquitin (Ub) activity (viscera), and lipid peroxidation (LPO; muscle) was observed under the combination of warming and PST exposure. The differential bioaccumulation and biomarker responses observed highlight the need to further understand the interactive effects between PST and abiotic stressors, to better estimate climate change impacts on HABs events, and to develop mitigation strategies to overcome the potential risks associated with seafood consumption.

Keywords: Paralytic shellfish toxin; warming; fish; seafood safety; ecotoxicological responses

Key Contribution: Warming conditions can promote higher PST accumulation in *S. aurata* juveniles. Moreover, the co-exposure of warming with PST affected animal condition as well as fish ecotoxicological responses, resulting in the inhibition of the antioxidant machinery and the enhancement of cellular damage. The present study provides novel and significant insights for a better understanding on toxin accumulation in fish species under climate change scenarios.

1. Introduction

Harmful algae blooms (HABs) naturally occur under favorable environmental conditions, leading to the proliferation and/or aggregation of microalgae species containing high levels of toxic compounds, i.e., marine biotoxins [1]. The geographic distribution of toxic algae species has been associated with changes in local or regional eutrophication conditions, or due to large-scale climatic changes [2]. Indeed, coastal eutrophication and extreme climate events, such as El Niño, may promote favorable growing conditions (i.e., nutrient enriched waters) for the occurrence of toxic algal blooms, and therefore increased HAB events [1,2]. HABs are a major concern for marine ecosystems, as they can translate in several toxicological effects to the marine species that ingest them, being particularly deleterious to individuals in early life stages [3,4]. Moreover, HAB events have a great impact on human health, due to the consumption of contaminated seafood [1]. Filter-feeding organisms, such as bivalves, feed toxic microalgae and accumulate toxins they produce. Recently, other taxonomic groups higher up in the food chain (e.g., predatory fish, cephalopods, birds, and mammals) have been also pointed out as important vectors of marine biotoxins. Yet, so far, little attention has been paid to the transfer and toxicological mechanisms of marine toxins in these "emerging vector species" [5,6].

Marine biotoxins can be classified according to their solubility (i.e., hydrophilic or lipophilic), as well as their toxicological mode of action (i.e., paralytic shellfish poisoning, PSP; amnesic shellfish poisoning, ASP; diarrheic shellfish poisoning, DSP; neurotoxic shellfish poisoning, NSP; and ciguatera fish poisoning, CFP) [7]. Among the hydrophilic biotoxins are paralytic shellfish toxins (PST) and amnesic shellfish toxins (AST), whereas diarrheic shellfish toxins (DST), neurotoxic shellfish toxins (NST), and ciguatoxins (CTX) are lipophilic compounds [8]. Paralytic shellfish toxins (PST), including saxitoxin and saxitoxin-related compounds (STXs), are potent neurotoxins mainly produced by marine dinoflagellates that cause PSP [1,8]. PST neurotoxicity is due to their high affinity to bind to voltage-gated sodium channels, inhibiting the passage of sodium ion nerve cell membranes, and thus blocking neuronal and muscular activities [4,8]. As PST toxicity differs according to the binding affinity of each compound [7], carbamate toxins, including saxitoxin (STX), neosaxitoxin (NEO), and gonyautoxins (GTX1 and GTX4) have been considered the most toxic PSTs, followed by their decarbamoyl derivatives (dcSTX, dcGTX, and dcNEO), whereas N-sulfocarbamoyl toxins (e.g., B1 (GTX5), B2 (GTX6), and C1 and C2) are usually associated with lower toxicity [7,8].

Over the past decades, HABs have increased in frequency, intensity, and geographic distribution, mainly due to the increase in seawater temperatures that favors microalgae growth [9]. Indeed, worldwide, climate change is increasing seawater surface temperature (SST), and this trend is expected to worsen over the next decades, with SSTs increasing up to 5 °C in some regions [2,10]. Yet, both direct and indirect impacts of climate change effects in marine ecosystems, especially in the food-web system, are still unclear [9]. Understanding the way and extent to which abiotic variables can affect the occurrence/toxicity of HABs in seafood species will make it possible to anticipate how climate change drivers affect marine species from both an ecological and a seafood safety perspective.

PST exposure [11] and climate change effects [12] are known to induce adverse effects on fish species, mainly in the antioxidant mechanism as a result of oxidative stress [13]. Induced oxidative stress may exert cytotoxic effects through the overproduction of reactive oxygen species (ROS), which are involved in cellular protective mechanisms, but at higher concentrations lead to deleterious effects in cellular proteins, lipids, and DNA [14]. Several biochemical assays can be used to evaluate the ecotoxicological responses induced by exposure to environmental contaminants and climate change stressors [12,14]. Within fish antioxidant machinery, catalase (CAT) and superoxide dismutase (SOD) are considered ROS scavengers with protective roles against ROS formation, while glutathione S-transferases (GST) plays a key role in organs' second phase detoxification [15]. In addition, heat shock proteins (HSP) are mainly associated with cellular redox changes by temperature, and lipid peroxidation (LPO) is the ultimate degradation product of cellular injury [15]. Yet, such an approach has not been employed to evaluate the ecotoxicological effects of PST under warming.

Within this context, the present study aims to assess, for the first time, the effect of seawater temperature regimes on PST (C1 and C2, dcSTX and B1) bioaccumulation and depuration mechanisms in juvenile fish, as well as its ecotoxicological responses, following five days of dietary exposure to these toxins. Gilthead seabream (*Sparus aurata*) was selected as the biological model, since it is a predatory fish with high commercial value, widely produced in coastal areas of the eastern Atlantic and Mediterranean Sea [16]. Blue mussels (*Mytilus galloprovincialis*) constitute a natural prey of fish species inhabiting the Mediterranean region, such as *S. aurata*, and this bivalve species is a primary vector of PST in coastal areas [17]. Therefore, naturally-contaminated mussels were used as feed to expose juvenile seabream to PST.

2. Results

2.1. Influence of Warming on PST Accumulation and Depuration

PST were detected after four days of exposure (regardless of temperature regime), with the highest concentration being found on day five at 24 °C (0.97 µg g^{-1} C1 and C2, 0.57 µg g^{-1} B1 and 0.09 µg g^{-1} dcSTX) (Figure 1). The PST profile was limited to C1 and C2, dcSTX and B1 toxins analogues, matching the toxin profile of contaminated mussels' hepatopancreas used as feed. C1, C2, and B1 toxins were the most abundant PST in seabream juvenile specimens (Figure 1A,C). Still, on day four, higher concentrations were observed for the B1 toxin (0.27 µg g^{-1}; Figure 1C), whereas on day 5, higher concentrations were observed for C1 and C2 (0.97 µg g^{-1}; Figure 1A). PST were not detected (levels below detection limit) during the depuration period (i.e., days 6–10), indicating a fast elimination rate in this fish species (Figure 1A–C).

On day four, higher temperatures (21 °C and 24 °C) significantly increased ($p < 0.05$) B1 toxin levels in seabream juveniles (Figure 1C), whereas on day five (i.e., maximum PST exposure), significantly higher concentrations of C1, C2, and B1 toxins were observed in fish exposed to the highest seawater temperature (i.e., 24 °C; $p < 0.05$; Figure 1A,C). In addition, warming significantly increased C1 and 2 concentration with time (day five > day four), while dcSTX concentration was significantly higher on day five, regardless of seawater temperature (Figure 1A,B). PST toxicity was calculated using the toxicity equivalency factors (TEFs) adopted for each toxin group [18]. Warming significantly increased ($p < 0.05$) PSP toxicity at the maximum exposure period, where the maximal toxicity of 275 ± 3 µg STX eq. kg^{-1} was reached on day five and 24 °C (Table 1). Moreover, PSP toxicity in seabream juveniles significantly increased from day four to day five at higher temperatures (21 °C and 24 °C; Table 1).

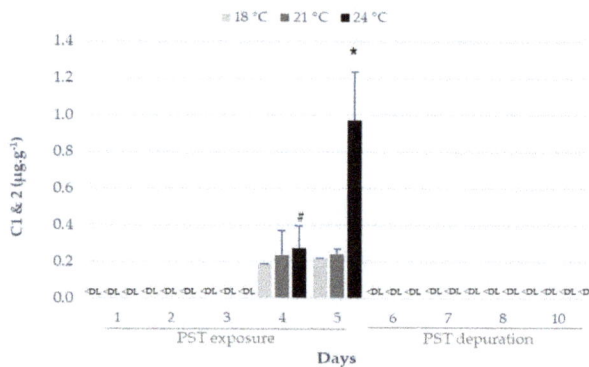

(A)

Figure 1. *Cont.*

(B)

(C)

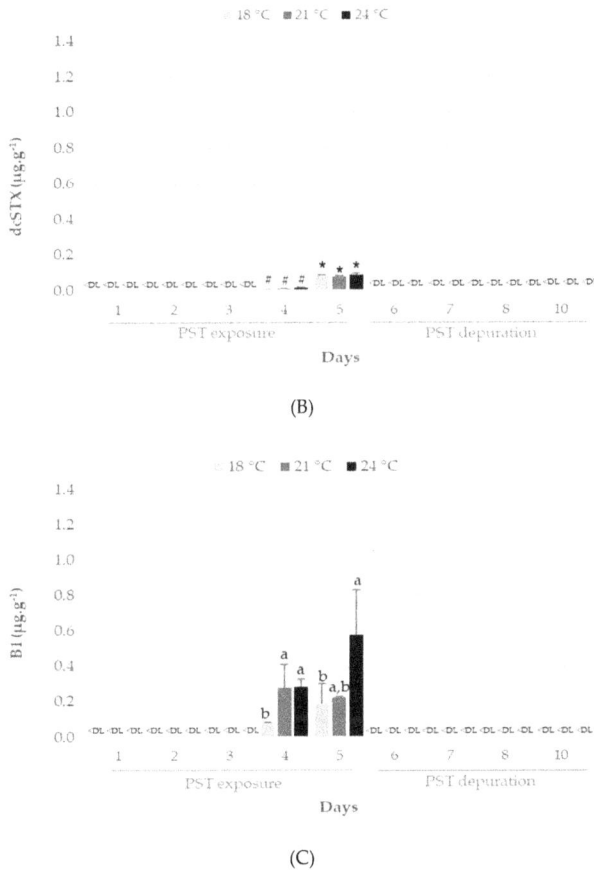

Figure 1. Effect of different temperature regimes (18 °C, 21 °C, and 24 °C) on the accumulation/depuration of paralytic shellfish toxins (PST) ($\mu g\ g^{-1}$) in *Sparus aurata*: (**A**) N-sulfocarbamoylgonyautoxin-1 and -2 (C1 and C2), (**B**) decarbamoylsaxitoxin (dcSTX), (**C**) gonyautoxin-1 (B1), during the experimental period. Results are expressed as mean ± SD ($n = 5$). Different letters (a, b, c) indicate significant differences ($p < 0.05$) between temperatures, whereas the symbols (*, #) indicate significant differences ($p < 0.05$) between days. <DL = below detection limit.

Table 1. Toxicity (μg STX eq. kg^{-1}) of *S. aurata* exposed via feed to PST at different temperatures and current EU limit [11] for paralytic shellfish poisoning (PSP) toxins.

Sampling Day	Temperature	Toxicity (μg STX eq. kg^{-1})	EC 853/2004[11]
Day 4	18 °C	74.7 ± 1.6	
	21 °C	59.1 ± 8.3 [#]	
	24 °C	67.1 ± 6.5 [#]	800 μg STX eq. kg^{-1}
Day 5	18 °C	113.4 + 17.2 [b]	
	21 °C	154.1 ± 1.2 [b,*]	
	24 °C	275.4 ± 3.0 [a,*]	

Different letters (a, b) indicate significant differences ($p < 0.05$) between temperature (18 °C, 21 °C, and 24 °C) for the same day, whereas the symbols (*, #) indicate significant differences ($p < 0.05$) between days for the same temperature. Results are expressed as mean ± SD ($n = 5$).

2.2. Influence of Warming and PST Exposure on Fish Ecotoxicological Responses

No mortality or changes in fish behavior were observed during the experiment. The combined effect of PST exposure and warming significantly decreased ($p < 0.05$) animal condition, as a significantly lower Fulton's condition index (K) was observed at 24 °C and on day five (PST exposure) (Figure 2). On the other hand, no significant differences ($p < 0.05$) were observed on day 10 (depuration), regardless of water temperature (Figure 2).

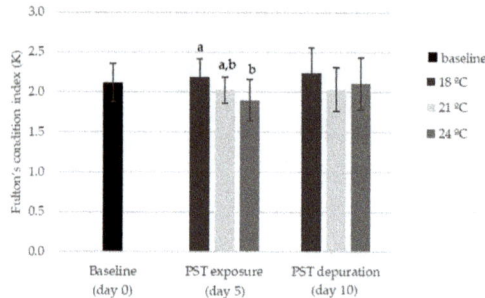

Figure 2. Fulton's condition index (K) in *S. aurata* before PST exposure (day 0), after five days of exposure (day 5) and after five days of depuration (day 10) at different temperatures (mean ± SD; $n = 5$). Different letters (a, b) indicate significant differences ($p < 0.05$) between treatments (18 °C, 21 °C, and 24 °C) for each day.

The levels of oxidative stress-related enzymes, including SOD, CAT, and GST activities, are presented in Figure 3. In fish viscera, warming (i.e., exposure to 21 °C and 24 °C) significantly inhibited ($p < 0.05$) SOD activity after PST exposure (day five), as well as during the depuration phase (day ten; $p < 0.05$; Figure 3A). Conversely, acclimation to warmer temperatures (before PST exposure, i.e., day zero) induced lower SOD activity in fish muscle ($p < 0.05$), such a trend was reversed after five days of concomitant exposure to PST, with SOD inhibition increasing in all treatments, regardless of temperature regime (Figure 3B). Moreover, SOD activity tended to decrease throughout time (i.e., day zero versus day ten) in the muscle of fish exposed to the lowest temperature regime (i.e., at 18 °C), but not in those exposed to warmer temperatures (Figure 3B). Warming (both temperatures) significantly reduced CAT activity in the viscera and muscle of fish, regardless of PST exposure ($p < 0.05$; Figure 2C,D), only except during the depuration period (day ten) in muscle ($p < 0.05$; Figure 3D). Noteworthy, throughout the experimental period, CAT activity decreased in the muscle of fish under the control temperature (18 °C), being significantly lower ($p < 0.05$) on day ten (PST depuration) compared to day zero (baseline) and day five (PST exposure; Figure 3D). At the beginning of the experiment (day zero, baseline), fish acclimated under warmer temperatures exhibited significantly lower GST activity ($p < 0.05$) in both tissues compared to those under the control temperature (Figure 3E,F). After five days of PST exposure (i.e., on day five), this trend was maintained in fish muscle (i.e., GST activity stayed significantly lower in fish exposed to warming, particularly to the highest temperature; Figure 3F) but not in the viscera, with PST exposure being responsible for a significant diminishment of GST activity in fish exposed to the control temperature (on both day five and day ten in viscera and on day ten in muscle; Figure 3E,F).

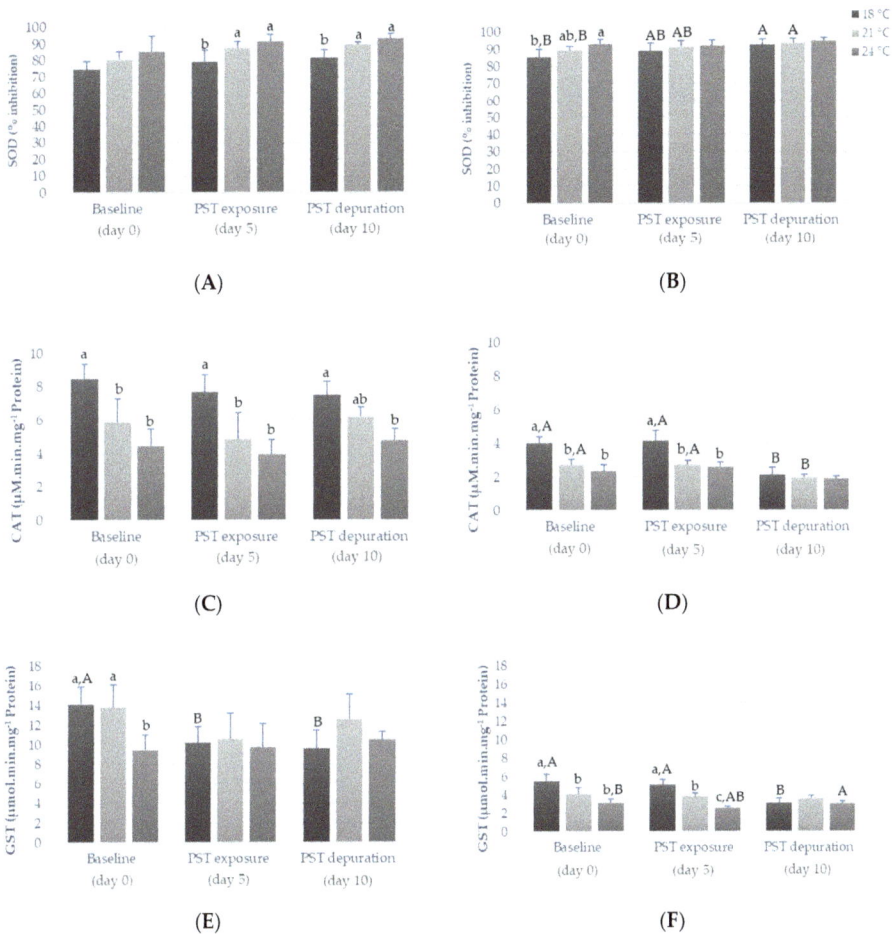

Figure 3. Anti-oxidant enzyme activities (SOD; CAT; GST) in the viscera (**A,C,E**) and muscle (**B,D,F**) of *S. aurata* before PST exposure (day 0), after five days of PST exposure (day 5) and after five days of depuration (day 10) at different temperatures (18 °C, 21 °C, and 24 °C). Results are expressed as mean ± SD (*n* = 5). Different letters (a, b, c) indicate significant differences between temperatures ($p < 0.05$), whereas (A, B) indicate significant differences between days. Abbreviations: CAT—catalase activity; SOD—superoxide dismutase inhibition; GST—glutathione S-transferase activity.

Matching the overall inhibition of antioxidant enzyme activities promoted by warmer temperatures and/or PST exposure, LPO (measured as malondialdehyde (MDA) concentration) gradually increased over time in fish exposed at 21 °C and 24 °C, being significantly higher ($p < 0.05$) compared to the values observed in fish under 18 °C after five days of PST exposure (muscle) as well as after the PST depuration period (day 10; viscera and muscle; Figure 4A,B).

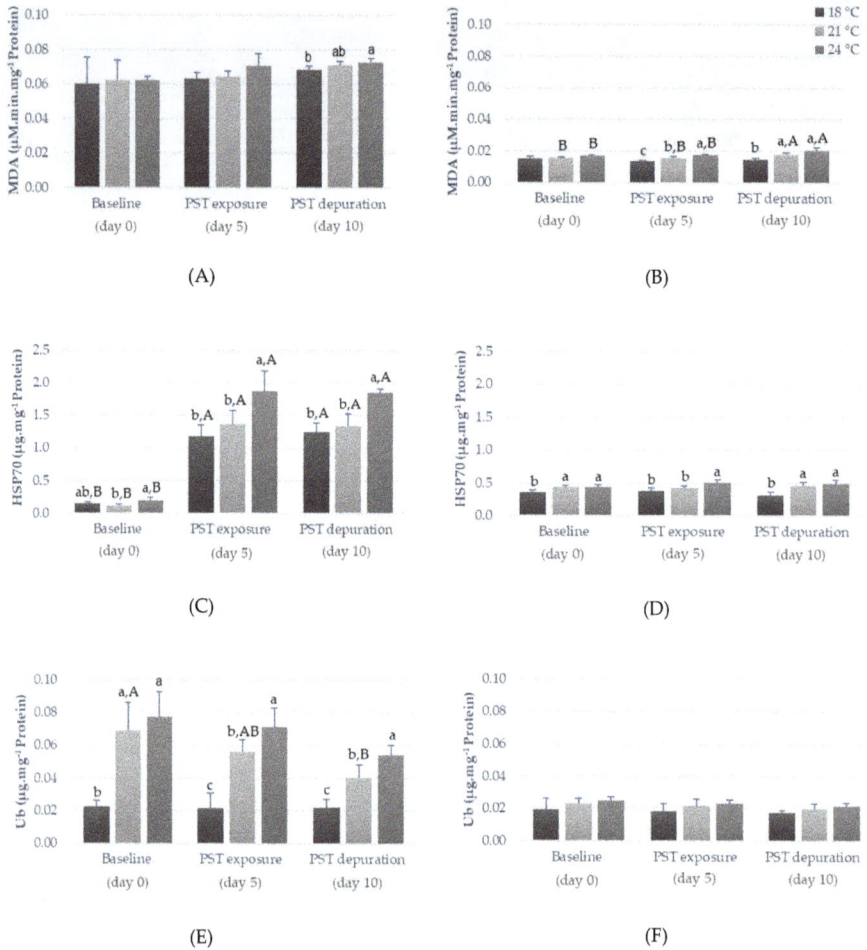

Figure 4. Lipid peroxidation (as MDA concentration), heat shock protein (HSP70) concentration, and ubiquitin concentration (Ub) in viscera (**A,C,E**) and muscle (**B,D,F**) of *S. aurata* before PST exposure (day 0), after five days of PST exposure (day 5) and after five days of depuration (day 10) at different temperatures (18 °C, 21 °C, and 24 °C). Results are expressed as mean ± SD (n = 5). Different letters (a, b, c) indicate significant differences between temperatures ($p < 0.05$), whereas (A, B) indicate significant differences between days. Abbreviations: MDA—malondialdehyde concentration.

Concerning heat shock response, HSP70 content in both fish tissues was significantly affected by temperature and PST exposure (Figure 4C,D). In fish viscera, PST exposure triggered a drastic increase in HSP70 proteins synthesis ($p < 0.05$), particularly at warmer temperatures (24 °C), a trend that was still observed even after the five days of the PST depuration period (Figure 3C). Conversely, in fish muscle, HSP70 levels did not seem to be significantly affected by PST exposure (i.e., no significant differences between days zero, five, and ten), whereas warmer temperatures, particularly 24 °C, increased the synthesis of these proteins for all sampling days ($p < 0.05$; Figure 4D). Warming (21 °C and 24 °C) significantly increased ($p < 0.05$) Ub protein synthesis in fish viscera, regardless of PST exposure, though a gradual decrease was observed throughout time in these two treatments ($p < 0.05$). Conversely, this tendency was not observed in the one simulating the control temperature (an increase between days zero, five, and ten was observed instead in fish exposed at 18 °C; Figure 4E). In comparison, fish muscle did not evidence significant differences in Ub contents (Figure 4F) nor in AChE activity (Figure 5).

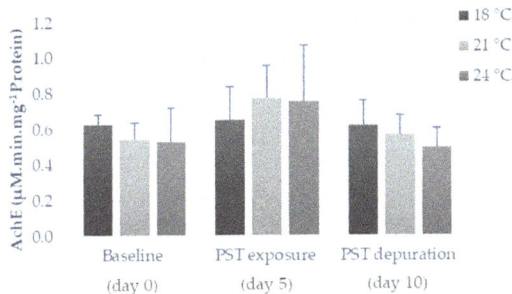

Figure 5. Acetylcholinesterase (AChE) activity in muscle tissue of *S. aurata* before PST exposure (day 0), after five days of PST exposure (day 5) and after five days of depuration (day 10) at different temperatures (18 °C, 21 °C, and 24 °C). Results are expressed as mean ± SD ($n = 5$).

3. Discussion

3.1. Influence of Warming on PST Accumulation and Depuration

In line with the present findings, low levels of toxins were observed in fish exposed to PST through feed, with toxin concentrations evidencing an increase with the exposure time (i.e., maximum concentration found by the end of the exposure period) [19,20]. Yet, contrasting with previous studies, PST profiles in fish (predators) are identical to the profiles of their prey (C1 and C2 > B1 > dcSTX). Only a few studies have focused on fish metabolism of PST. However, the differences previously reported in the toxin profiles of prey and predators suggest that PST biotransformation may also take place [19,21]. Nevertheless, the low levels detected, associated with identical elimination rates during uptake and depuration, may explain the absence of PST metabolization [22]. Several studies show that viscera are the primary organ for PST accumulation, but have also higher detoxification rates (excretion), which can be effectively accelerated in juvenile specimens that present rapid growth, and therefore faster metabolism [22].

In agreement with the present results, Costa et al. [19] and Kwong et al. [20] reported high toxin elimination in *Diplodus sargus* (B1 and dcSTX) and *Acanthopagrus schlegeli* (C1 and C2), suggesting that PST can be easily excreted by renal processes [23]. Interestingly, PSTs were not detected after the first 24 h of depuration. In contrast, previous reports showed decreased toxin concentrations during the first five days of depuration [19,21]. Such differences may be explained by different toxin profiles and model fish species. It is known that C2 and C1 toxin analogues, which were the predominant ones in our study, are less stable and easily undergo enzymatic hydrolysis, being rapidly eliminated via urine [19]. Moreover, warmer water temperatures lead to higher metabolic rates associated with the increase in fish energetic demands [24] and, consequently, higher excretion rates of the more soluble toxin analogues [22].

To the authors' best knowledge, so far, the effect of warming on PST accumulation/depuration has only been assessed in bivalve species, i.e., oysters (*Crassostrea gigas* and *Saccostrea glomerata* [25]), sea scallops (*Placopecten magellanicus* [26]) and mussels (*Mytilus edulis* [26] and *Mytilus galloprovincialis* [27]), therefore hampering adequate comparisons of the present data (i.e., concerning a fish model species) with previous studies. However, while in *M. edulis* and *P. magellanicus* the effect of temperature on PST uptake was unclear [26], warmer temperatures significantly decreased PST concentrations in *S. glomerata*, diploid *C. gigas* [25], and *M. galloprovincialis* [27]. Contrarily, a previous study showed PST accumulation in the fish muscle of *Geophagus brasiliensis* during HABs, with slightly higher PST concentration in summer compared to spring and autumn, in a Brazilian reservoir [28]. Several studies demonstrated that warmer temperatures enhance organic compound accumulation in fish species (e.g., Hg in *Dicentrarchus labrax* [29], triclosan in *Diplodus sargus* [24]) via feed ingestion, as a result of enhanced fish metabolism and, therefore, increased feeding rates. Still, increased metabolic rates can also translate into increased compound metabolization and/or excretion [28], which may explain the present results. Despite the increased accumulation of PST in seabream at warmer temperatures, toxin concentrations remained below the current safety limits established for human consumption (800 μg STX eq. kg^{-1}) [18]. Nevertheless, these limits were established to protect consumers [18], and therefore the potential toxic effect on fish welfare can be under or overestimated. Indeed, in terms of fish welfare it is worthwhile highlighting the ecotoxicological responses observed in juvenile gilthead seabream fish exposed to PST at warmer temperatures (e.g., decreased animal fitness). Yet, particularly noteworthy was the 1.4-fold increase in seabream toxicity with warming (24 °C), representing a 20% increase in PSP toxicity. These results strongly suggest that the higher toxin accumulation levels might be exacerbated if temperatures continue to increase to levels projected by the Intergovernmental Panel on Climate Change (IPCC) for the worst-case scenario. It is known that several bivalve species can convert N-sulfocarbamoyl toxins to their corresponding carbamate toxin (more toxic) under conditions of a high temperature and low pH [23]. Yet, such a process of conversion and/or metabolized in fish is still unclear and could not be detected in the current study.

Furthermore, in line with previous studies [30], the present results suggest that differences would be expected with adult seabream specimens. It is known that adult fish have higher rates of feed ingestion, meaning the ingestion of higher amounts of feed and, consequently, the ingestion of higher toxin concentrations. In addition, evidence of toxin biotransformation (the conversion of less potent and less stable toxins into more potent and stable ones) during digestion can occur by enzymes in adult fish, as well as toxin distribution through extravascular fluids to other organs revealing PST bioaccumulation within the food chain [30].

3.2. Influence of Warming and PST Exposure on Fish Biochemical Responses

In agreement with previous studies, an increased oxidative stress response was observed in both tissues due to warming and PST exposure in *S. aurata* [12,24,28,31,32]. Aquatic organisms' antioxidant mechanisms are complex, and exposure to stressful environmental conditions can induce or inhibit antioxidant enzymes [28]. Generally, increased SOD and CAT activities are linked with warmer seawater temperatures, due to the enhancement of an organism's metabolism [12,24,32], whereas SOD and CAT activities inhibition were observed after toxin and metal exposure [31–33]. The present results show that the combined exposure to PST and warming resulted in lower SOD activity in viscera in the beginning of the exposure trial, while in muscle, SOD activity was inhibited by warmer temperatures, regardless of PST exposure. It is known that the liver is the main organ for PST accumulation, being responsible for toxin biotransformation, redistribution to other tissues, and elimination [19]. Yet, both warming and PST exposure reduced CAT activity in viscera and muscle. A similar pattern was also found with xenobiotic compounds (e.g., endosulfan in *Channa punctatus* [34], MeHg in *Dicentrarchus labrax* [32], triclosan in *Diplodus sargus* [24], and STX in *Hoplias malabaricus* [35]). PST exposure may result in an intensive formation of reactive oxygen species (ROS), and such excessive substrate production (superoxide anion) may inhibit CAT activity [35]. As for GST, the activity inhibition in fish exposed to warmer temperatures (viscera and muscle) and PST (muscle) is in contradiction with previous studies carried out with MeHg [32] and other fish species, such as Atlantic salmon [36] and *Hoplias malabaricus* [35], where GST activity increased under stressful conditions and/or as a response to CAT activity inhibition [24,28,32,36]. PST metabolization mainly occurs in fish liver, where phase I and phase II reactions for the biotransformation of xenobiotics take place [20]. It has been hypothesized that C1, C2, and B1 analogues can enter directly phase II of biotransformation, yet so far GST induction has mainly been reported in fish exposed to PST carbamate analogues [20,21]. Furthermore, enzyme denaturation or cells' inability to synthetize enzymes can occur when temperatures exceed threshold values [12]. On the other hand, GST activity inhibition indicates that the N-sulfocarbamoyl toxins (C1, C2, and B1) were highly hydrophilic and easily eliminated, not being necessary to promote PST analogue excretion through the conjugation of toxins with reduced glutathione (GHS; GST catalyze) [21]. Antioxidant enzymes are strongly species-dependent, as each species has different thermal tolerance limits and enzyme baseline levels [12,37]. Yet, the distinct trends observed show that increased temperatures affect fish antioxidant responses to PST exposure, in a tissue- and biomarker-specific way (e.g., enhanced SOD activity was observed in viscera, but not in muscle, and inhibited GST activity in muscle, but not in viscera). Interestingly, PST exposure strongly affected fish antioxidant machinery in muscle at 18 °C, since significantly lower SOD, CAT, and GST activities were observed after the PST depuration period, compared to the baseline, corroborating the fact that antioxidant enzyme activities are time-dependent [37].

The increase in LPO under warming and PST exposure indicate that membrane damage occurred over time (i.e., overall, higher values at the end of the trial compared to baseline values), particularly at warmer temperatures (possibly due to SOD, CAT, and GST inhibition), further corroborating the time-dependency of cells' antioxidant scavengers and indicating that these enzymes were not able to totally prevent the oxidative damage induced by ROS, potentially leading to cell death [32]. The significantly higher MDA concentration observed in muscle after PST depuration at higher temperatures suggests that the combined effect of warming and PST exposure may lead to irreversible cell damage.

The increase in HSP70 is generally associated with rising temperatures, as well as with exposure to environmental contaminants [32,37]. Generally, HSP content gradually increases until reaching a maximum level and then it starts to decrease as thermal stress becomes more severe and protein synthesis mechanisms are led to exhaustion [12]. Noteworthy, HSP synthesis may be influenced by the baseline contents of each species and/or tissues, and by the synergistic effects of contaminants [32], explaining the significant increase in HSP content in viscera after PST exposure compared to non-exposed fish. In addition, the increased Ub levels in viscera after warming and PST exposure indicate that the synthesis of these proteins was triggered in response to stressful conditions, most likely as a result of an increased need for chaperoning and degradation by the proteasome of protein anomalies [37]. In what concerns muscular AChE, Clemente et al. [28] reported a significant increase after PST exposure at higher temperatures (summer). However, in the present work, both warming and PST exposure did not seem to affect AChE activity in fish muscle.

4. Conclusions

The present study provides evidence that increased seawater temperatures facilitate PST bioaccumulation in juvenile seabream specimens, despite the possibility for these toxins to be rapidly depurated. Although warming promoted higher toxin accumulation by fish (1.4-fold increase), PSP toxicity levels remained below the current safety limits established for human consumption.

In terms of fish ecotoxicological responses, the co-exposure of warming with PST decreased animal fitness (K), and affected the biomarker responses of fish tissues, resulting in the inhibition of antioxidant scavengers (SOD, CAT, and GST) as well as in the enhancement of biomarkers involved in lipid (LPO) and protein (HSP70 and Ub) damage in cells. Yet, the impairment of fish antioxidant machinery under warming and PST exposure, alone or in co-exposure, suggests that ecotoxicological responses can only prevent oxidative stress to some extent, inducing cell damage, health problems and ultimately, fish mortality.

The different PST accumulation observed in fish exposed to warming conditions highlights the need to consider interactions between multiple stressors, especially linked with climate change scenarios (i.e., HABs, warming, and acidification), in future studies on toxin accumulation and elimination in commercial marine species, as well as for ecotoxicological responses. Such studies will allow the collection of more realistic information on the potential effects of climate change-related stressors on HABs toxicity potentially causing the impossibility to trade commercially valuable fish species.

5. Materials and Methods

5.1. Preparation of PST Contaminated Diet

Naturally contaminated and non-contaminated mussels (*Mytilus galloprovincialis*) were used to expose fish to PST through their diet. Contaminated *M. galloprovincialis* were collected in Aveiro Lagoon, NW Portuguese coast, during a bloom of *Gymnodinium catenatum* in late 2016. The presence of PST was confirmed by liquid chromatography, as previously described by Costa et al. [19]. Toxin composition included carbamate (STX, GTX2, and GTX3), N-sulfocarbamoyl (C1, C2, and B1) and decarbamoyl analogues (dcGTX2, dcGTX3, and dcSTX) (Table 2), and the PST toxicity measured was 27,232 µg STX eq. kg^{-1}. Mussels hepatopancreas were dissected and freeze-dried at −50 °C, 10^{-1} atm of vacuum pressure, for 48 h (Power Dry 150 LL3000, Heto, Czech Republic), homogenized, and kept at −20 °C prior to the feeding experiments.

Table 2. Toxin profile (mg kg^{-1}) in mussels (*M. galloprovincialis*) hepatopancreas given as food to gilthead seabream (*S. aurata*).

Toxins Analogues	DW	WW
dcGTX2 and dcGTX3	4.6	1.1
C1 and C2	248.9	60.8
dcSTX	50.4	12.3
GTX2 and GTX3	1.5	0.4
B1	156.1	38.1
STX	0.6	0.1
Total Toxicity (mg STX eq. kg^{-1})	111.5	27.23

DW—dry weight; WW—wet weight. dcGTX2 and dcGTX3 (decarbamoylgonyautoxin-2 and -3); C1 and C2 (N-sulfocarbamoylgonyautoxin-2 and -3); dcSTX (decarbamoylsaxitoxin); GTX2 and GTX3 (gonyautoxin-2 and -3); B1 (gonyautoxin-1); STX (saxitoxin).

5.2. Experimental Design and Biological Sampling

Juvenile specimens of *S. aurata* reared at the aquaculture pilot station of the Portuguese Institute for the Sea and Atmosphere (EPPO-IPMA, Olhão, Portugal), were maintained in 24 rectangular glass tanks (~50 L) in Guia Marine Laboratory (MARE-FCUL, Cascais, Portugal, with an independent water recirculation system (RAS), temperature and pH control (Profilux 3.1N, GHL, Germany), refrigeration system (Frimar, Fernando Ribeiro Lda, Portugal), protein skimmers (Reef SkimPro, TMC Iberia, Portugal), UV disinfection (Vecton 300, TMC Iberia, Portugal), and biological filtration (model FSBF 1500, TMC Iberia, Portugal). Seawater parameters were controlled daily through seawater renewal (25%) and by colorimetric tests (Tropic Marin, Montague, CA, USA). Ammonia and nitrites were kept below detectable levels, while nitrates were kept below 2.0 mg L^{-1}. Seabream specimens were acclimated for 15 days in aerated seawater (dissolved O$_2$ > 5 mg L^{-1}) at 18 ± 0.5 °C, pH 8.0 ± 0.1 units, 35 ± 1.0 ‰ salinity, 12:12 h photoperiod and fed with 7% of the average body weight (b.w.), with a commercial fish diet manufactured by SPAROS, Lda (Olhão, Portugal). Detailed feed nutritional composition can be consulted in Table S1. Five days before initiating PST exposure, seawater temperature was slowly adjusted (1.0 ± 0.5 °C per day), until it reached 21 °C and 24 °C in the tanks, simulating warming conditions [10] and heat wave conditions [10,38], respectively. During this period, the commercial fish diet used to feed fish was replaced by lyophilized hepatopancreas of non-contaminated mussels (amount equivalent to 7.6% of fish b.w. with PST < DL), to allow fish to adapt to this new type of food. Three scenarios were carried out (*n* = 18 animals per replicate tank of treatment, i.e., total of 144 animals per treatment), simulating the current temperature conditions used in seabream rearing (18 °C), an increase in average seawater temperature simulating the warming conditions projected by the IPCC in the Mediterranean region (ΔT °C = +3 °C; RCP 8.5, IPCC, 2014), and seawater temperature increase simulating a heat wave event (ΔT °C = +6 °C; [39]) (Figure 6). During the five days of PST exposure, seabream juveniles were daily fed with PST contaminated mussels (lyophilized hepatopancreas; 7.6% b.w.; toxins' profile presented in Table 2), and subsequently fed again with non-contaminated mussels (7.6% b.w.) during the five days of the depuration phase. During acclimation and the exposure trial, seawater abiotic parameters were checked daily and adjusted whenever needed.

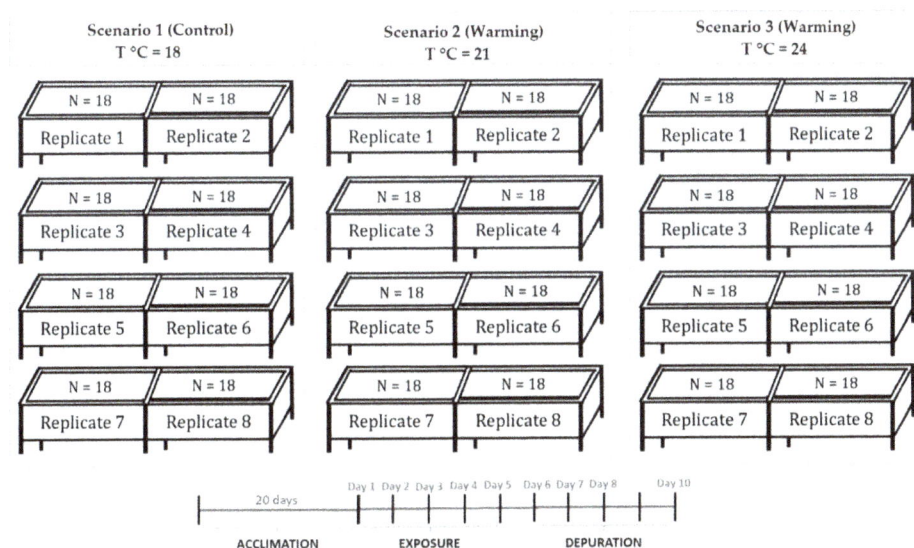

Figure 6. Experimental setup.

For toxin extraction and quantification, 45 individuals were randomly collected 2 h after feeding, at days one, two, three, four, five (PST exposure), six, seven, eight, and ten (PST depuration). Fish were randomly collected from each treatment and euthanized by immersion in an overdosed MS222 solution (2000 mg L^{-1}; Sigma-Aldrich, St. Louis, MO, USA) buffered with sodium bicarbonate (1 g of NaHCO$_3$ to 1 g of MS222 to 1 L of seawater). In each temperature, 15 specimens were collected (n = five fish per replicate; three replicates) for each sampling day. Euthanized fish were measured (total length, TL, and weight, W) and whole body (without head) and immediately frozen at −20 °C until further analysis. For enzymatic and protein quantification assays, 15 seabream juveniles were randomly collected from each temperature, euthanized, and measured, at day zero (before PST exposure), day five (maximum PST exposure), and day ten (final day of depuration period). Fish were carefully dissected, and fish muscle and viscera tissues (i.e., liver, pancreas, and intestines) were collected and immediately frozen at −80 °C until further analysis. Details regarding fish biometry can be consulted in Table S2.

5.3. Toxins Extraction and Quantification

Toxins from whole fish homogenate were heat-extracted in 1% acetic acid, vortexed, and centrifuged (15,000× *g*) for 10 min. Extracts followed a solid-phase extraction (SPE) with an octadecyl bonded phase silica (Supelclean LC-18 SPE cartridge, 3 mL, Supelco, Bellefonte, PA, USA). Periodate and peroxide oxidations of PST were carried out and toxins were immediately quantified by high performance liquid chromatography with fluorescence detection (HPLC-FLD), based on the precolumn oxidation method developed by Lawrence and Niedzwiadek (2001) [39]. The HPLC-FLD equipment consisted of a Hewlett–Packard/Agilent Model 1290 Infinity quaternary pump, autosampler, column oven, and Model 1260 Infinity fluorescence detector. PST oxidation products were separated using a reverse-phase Supelcosil LC-18, 15 × 4.6, 5 μm column (Supelco, Bellefonte, PA, USA). The mobile phase gradient consisted of 0–5% B (0.1 M ammonium formate in 5% acetonitrile, pH 6) in the first 5 min, 5–70% B for the next 4 min, and back to 0% B in the next 2 min. Then, 100% mobile phase A (0.1 M ammonium formate, pH 6) was used for 3 min before the next injection. The flow rate was 1 mL min^{-1} and the detection wavelength set to 340 nm for excitation and 395 nm for emission. Instrumental limits of detection (S/N = 3) were 11 ng g^{-1} dcSTX, 12 ng g^{-1} STX, 12 ng g^{-1} B1, 19 ng g^{-1} for dcGTX2 and dcGTX3, and GTX2 and GTX3, 34 ng g^{-1} C1 and C2. Working standard solutions for calibration curves were prepared by the dilution of PST stock solutions in PST-free cleaned-up fish tissue extract. Certified calibration solutions for PST were purchased from the Certified Reference Materials Program of the Institute for Marine Biosciences, National Research Council, Canada (STX-e, NEO-b, GTX2-b and GTX3-b, GTX1-b and GTX4-b, dcSTX, dcGTX2 and dcGTX3, GTX5-b (B1), C1 and C2, and dcNEO-b).

5.4. Biochemical Assays

Fish tissues (muscle and viscera) were homogenized in ice-cold conditions with 1.5 mL of phosphate buffered saline (PBS; 140 mM NaCl, 3 mM KCl, 10 mM KH$_2$PO$_4$, pH = 7.40 ± 0.02; reagents from Sigma-Aldrich, Steinheim, Germany), using an Ultra-Turrax® device (T25 digital, Ika, Germany) and centrifuged in 2 mL microtubes for 15 min at 10.000× *g* and 4 °C. Then, the supernatants were transferred to new microtubes, immediately frozen, and kept at −80 °C until further analysis. Seven molecular biomarkers were selected to assess the biological responses to PST exposure and warming at the tissue level. A summary of the selected biomarkers is presented in Table 3, with reference to the different methodologies used (further details regarding these methodologies have been previously described by Madeira et al. [12], Maulvault et al. [24], and Maulvault et al. [32]). All biochemical analyses were performed in triplicate and using reagents of pro analysis grade or higher. Total protein levels were quantified in each sample in order to enable the subsequent normalization of each biomarker (i.e., given in mg of protein; methodology based on the Bradford assay [40]). All methodologies were adapted to 96-well microplates, as previously reported by Maulvault et al. [33].

Table 3. Summary of selected molecular biomarkers and the corresponding methodologies used.

Molecular Biomarker	Ecotoxicological Response	Methodology Used	References
Superoxide dismutase (SOD)	Oxidative stress	Enzymatic assay	[12,24,32]
Catalase (CAT)	Oxidative stress	Enzymatic assay	[12,24,32]
Glutathione S-transferase (GST)	Oxidative stress and xenobiotic detoxification phase II	Enzymatic assay	[12,24,32]
Heat shock response (HSP70)	Chaperoning, heat shock response	Indirect ELISA	[24,32]
Ubiquitin (Ub)	Protein degradation and DNA repair	Direct ELISA	[24,32]
Lipid peroxidation (LPO)	Oxidative stress and cellular damage	Thiobarbituric acid reactive substances (TBARS) method	[12,24,32]
Acetylcholinesterase (AChE)	Neurotoxicity	Enzymatic assay	[24,32]

5.5. Animal Fitness Index (Fulton's K index)

The Fulton's K index was directly calculated from the biometric data to determine fish condition, according to the formula,

$$K = 100 \times (W/TL^3),$$ (1)

where W is the total wet weight (g) and TL is total length (cm).

5.6. Statistical Analysis

Results were expressed as mean values ± standard deviation (SD). ANOVA assumptions of normality and homoscedasticity were tested through Kolmogorov–Smirnov and Levene tests, respectively. Data were log-transformed or square rooted transformed, whenever at least one of the ANOVA assumptions was not verified. To evaluate the presence of significant differences between whole organism PST accumulation and temperature, one-way ANOVA analysis was performed. In terms of biochemical biomarkers (CAT, SOD, GST, LPO, HSP70, Ub, and AChE) and fish condition

Toxins **2019**, *11*, 408

(W, TL, and K), a two-way ANOVA was carried out, using tissue (viscera and muscle), temperature (i.e., 18 °C, 21 °C, and 24 °C) and treatment (Baseline (day zero), PST exposure (day five), and PST depuration (day ten)) as variables. Subsequently, post-hoc Tukey HSD tests were performed. Potential correlations between biomarker levels and the animal fitness index (Fulton's K index) were performed by means of the Pearson's correlation coefficient. Statistical analyses were performed at a significance level of 0.05, using STATISTICA™ software (Version 7.0, StatSoft Inc., Tulsa, OK, USA).

Supplementary Materials: The following are available online at http://www.mdpi.com/2072-6651/11/7/408/s1, Table S1: Commercial feed WIN Fast composition, by SPAROS, Lda (Olhão, Portugal), Table S2: Total length (TL; cm) and weight (W; g) of sampled specimens of S. aurata (mean ± standard deviation; $n = 15$) during the experiment.

Author Contributions: P.R.C. and A.M. conceived the project. P.P.-F., P.R.C. and A.M. designed the experiment. V.B. and M.S. carried out the exposure experiment. V.B. and P.A. performed the instrumental analysis. V.B., A.L.M., P.A., P.P.-F., P.R.C. and A.M. were involved in data analysis and interpretation. All authors were involved in the preparation of the manuscript.

Funding: The research leading to these results has received funding from the European Union H2020-BG-2014-2015/H2020-BG-2015-2 under CERES project (grant agreement n° 678193). The Portuguese Foundation for Science and Technology (FCT) supported the contract of AM in the framework of the IF2014 and CEECIND 2017 programs (IF/00253/2014 and CEECIND/01739/2017, respectively). This work contributes to project UID/Multi/04326/2019 from the Portuguese Foundation for Science and Technology (FCT).

Acknowledgments: The technical support of EPPO, IPMA, staff was essential to supply juvenile seabream for the experiments. The technical support of IPMA biotoxins monitoring staff was also essential to ensure the collection of contaminated mussel samples for the trials. Guia Maritime Laboratory, Faculty of Sciences, Lisbon University, for supporting the experiment at their premises.

Conflicts of Interest: The authors declare no conflict of interest.

References

1. Van Dolah, F.M. Marine Algal Toxins: Origins, Health Effects, and Their Increased Occurrence. *Environ. Health Perspect.* **2000**, *108*, 133–141. [CrossRef] [PubMed]
2. Marques, A.; Nunes, M.L.; Moore, S.K.; Strom, M.S. Climate change and seafood safety: Human health implications. *Food Res. Int.* **2010**, *43*, 1766–1779. [CrossRef]
3. Costa, P.R.; Botelho, M.J.; Lefebvre, K.A. Characterization of paralytic shellfish toxins in seawater and sardines (*Sardina pilchardus*) during blooms of *Gymnodinium catenatum*. *Hydrobiologia* **2010**, *655*, 89–97. [CrossRef]
4. Mincarelli, L.F.; Paula, J.R.; Pousão-Ferreira, P.; Rosa, R.; Costa, P.R. Effects of acute waterborne exposure to harmful algal toxin domoic acid on foraging and swimming behaviours of fish early stages. *Toxicon* **2018**, *156*, 66–71. [CrossRef] [PubMed]
5. Deeds, J.R.; Landsberg, J.H.; Etheridge, S.M.; Pitcher, G.C.; Longan, S.W. Non-traditional vectors for paralytic shellfish poisoning. *Mar. Drugs* **2008**, *6*, 308–348. [CrossRef] [PubMed]
6. Costa, P.R.; Costa, S.T.; Braga, A.C.; Rodrigues, S.M.; Vale, P. Relevance and challenges in monitoring marine biotoxins in non-bivalve vectors. *Food Control* **2017**, *76*, 24–33. [CrossRef]
7. Etheridge, S.M. Paralytic shellfish poisoning: Seafood safety and human health perspectives. *Toxicon* **2010**, *56*, 108–122. [CrossRef] [PubMed]
8. Visciano, P.; Schirone, M.; Berti, M.; Milandri, A.; Tofalo, R.; Suzzi, G. Marine Biotoxins: Occurrence, Toxicity, Regulatory Limits and Reference Methods. *Front. Microbiol.* **2016**, *7*, 1051. [CrossRef] [PubMed]
9. Gobler, C.J.; Doherty, O.M.; Hattenrath-Lehmann, T.K.; Griffith, A.W.; Kang, Y.; Litaker, R.W. Ocean warming since 1982 has expanded the niche of toxic algal blooms in the North Atlantic and North Pacific oceans. *Proc. Natl. Acad. Sci. USA* **2017**, *114*, 4975–4980. [CrossRef]
10. IPCC. *Climate Change 2014: Impacts, Adaptation, and Vulnerability. Part A: Global and Sectoral Aspects. Contribution of Working Group II to the Fifth Assessment Report of the Intergovernmental Panel on Climate Change*; Field, C.B., Barros, V.R., Dokken, D.J., Mach, K.J., Mastrandrea, M.D., Bilir, T.E., Chatterjee, M., Ebi, K.L., Estrada, Y.O., Genova, R.C., et al., Eds.; Cambridge University Press: Cambridge, UK; New York, NY, USA, 2014; 1132p.
11. Hong, H.Z.; Lam, P.K.; Hsieh, D.P. Interactions of paralytic shellfish toxins with xenobiotic-metabolizing and antioxidant enzymes in rodents. *Toxicon* **2003**, *42*, 425–431. [CrossRef]

12. Madeira, D.; Vinagre, C.; Diniz, M.S. Are fish in hot water? Effects of warming on oxidative stress metabolism in the commercial species *Sparus aurata*. *Ecol. Indic.* **2016**, *63*, 324–331. [CrossRef]
13. Perreault, F.; Matias, M.S.; Melegari, S.P.; Pinto, C.R.; Creppy, E.E.; Popovic, R.; Matias, W.G. Investigation of animal and algal bioassays for reliable saxitoxin ecotoxicity and cytotoxicity risk evaluation. *Ecotoxicol. Environ. Saf.* **2011**, *74*, 1021–1026. [CrossRef] [PubMed]
14. Ballesteros, M.L.; Wunderlin, D.A.; Bistoni, M.A. Oxidative stress responses in different organs of *Jenynsia multidentata* exposed to endosulfan. *Ecotoxicol. Environ. Saf.* **2009**, *72*, 199–205. [CrossRef] [PubMed]
15. Madeira, D.; Narciso, L.; Cabral, H.N.; Vinagre, C.; Diniz, M.S. Influence of temperature in thermal and oxidative stress responses in estuarine fish. *Comp. Biochem. Physiol. A* **2013**, *166*, 237–243. [CrossRef] [PubMed]
16. FAO. Cultured Aquatic Species Information Programme: *Sparus aurata*. In *Cultured Aquatic Species Information Programme*; Colloca, F., Cerasi, S., Eds.; FAO Fisheries and Aquaculture Department: Rome, Italy, 2005.
17. Šegvić-Bubić, T.; Grubišić, L.; Karaman, N.; Tičina, V.; Jelavić, K.M.; Katavić, I. Damages on mussel farms potentially caused by fish predation—Self-service on the ropes? *Aquaculture* **2011**, *319*, 497–504. [CrossRef]
18. EFSA. Scientific Opinion of the Panel on Contaminants in the Food Chain on a request from the European Commission on Marine Biotoxins in Shellfish—Saxitoxin Group. *EFSA J.* **2009**, *1019*, 1–76.
19. Costa, P.R.; Lage, S.; Barata, M.; Pousão-Ferreira, P. Uptake, transformation, and elimination kinetics of paralytic shellfish toxins in white seabream (*Diplodus sargus*). *Mar. Biol.* **2011**, *158*, 2805–2811. [CrossRef]
20. Kwong, R.W.M.; Wang, W.X.; Lam, P.K.S.; Yu, P.K.N. The uptake, distribution and elimination of paralytic shellfish toxins in mussels and fish exposed to toxic dinoflagellates. *Aquat. Toxicol.* **2006**, *80*, 82–91. [CrossRef]
21. Costa, P.R.; Pereira, P.; Guilherme, S.; Barata, M.; Nicolau, L.; Santos, M.A.; Pacheco, M.; Pousão-Ferreira, P. Biotransformation modulation and genotoxicity in white seabream upon exposure to paralytic shellfish toxins produced by *Gymnodinium catenatum*. *Aquat. Toxicol.* **2012**, *106–107*, 42–47. [CrossRef]
22. Costa, P.R. Impact and effects of paralytic shellfish poisoning toxins derived from harmful algal blooms to marine fish. *Fish* **2016**, *17*, 226–248. [CrossRef]
23. Bricelj, V.M.; Shumway, S.E. Paralytic shellfish toxins in bivalve mollusks: Occurrence, transfer kinetics and biotransformation. *Rev. Fish. Sci.* **1998**, *6*, 315–383. [CrossRef]
24. Maulvault, A.L.; Camacho, C.; Barbosa, V.; Alves, R.; Anacleto, P.; Cunha, S.C.; Fernandes, J.O.; Pousão-Ferreira, P.; Paula, J.R.; Rosa, R.; et al. Bioaccumulation and ecotoxicological responses of juvenile white seabream (*Diplodus sargus*) exposed to triclosan, warming and acidification. *Environ. Pollut.* **2019**, *245*, 427–442. [CrossRef] [PubMed]
25. Farrell, H.; Seebacher, F.; O'Connor, W.; Zammit, A.; Harwood, D.T.; Murray, S. Warm temperature acclimation impacts metabolism of paralytic shellfish toxins from *Alexandrium minutum* in commercial oysters. *Glob. Chang. Biol.* **2015**, *21*, 3402–3413. [CrossRef] [PubMed]
26. Mazur, M. The Effect of Temperature on Paralytic Shellfish Toxin Uptake by Blue Mussels (*Mytilus Edulis*) and Sea Scallops (*Placopecten magellanicus*). Honors College Paper 216. 2015. Available online: http://digitalcommons.library.umaine.edu/honors/216 (accessed on 23 May 2019).
27. Braga, A.C.; Camacho, C.; Marques, A.; Gago-Martínez, A.; Pacheco, M.; Costa, P.R. Combined effects of warming and acidification on accumulation and elimination dynamics of paralytic shellfish toxins in mussels *Mytilus galloprovincialis*. *Environ. Res.* **2018**, *164*, 647–654. [CrossRef] [PubMed]
28. Clemente, Z.; Busato, R.H.; Ribeiro, C.A.O.; Cestaric, M.M.; Ramsdorf, W.A.; Magalhães, V.F.; Wosiack, A.C.; Silva de Assis, H.C. Analyses of paralytic shellfish toxins and biomarkers in a southern Brazilian reservoir. *Toxicon* **2010**, *55*, 396–406. [CrossRef] [PubMed]
29. Maulvault, A.L.; Custodio, C.; Anacleto, P.; Repolho, T.; Pousão, P.; Nunes, M.L.; Diniz, M.; Rosa, R.; Marques, M. Bioaccumulation and elimination of mercury in juvenile seabass (*Dicentrarchus labrax*) in a warmer environment. *Environ. Res.* **2016**, *149*, 77–85. [CrossRef] [PubMed]
30. Bakke, M.J.; Horsberg, T.E. Kinetic properties of saxitoxin in Atlantic salmon (Salmo salar) and Atlantic cod (Gadus morhua). *Comp. Biochem. Physiol. Part C Toxicol. Pharmacol.* **2010**, *152*, 444–450. [CrossRef]
31. Silva de Assis, H.C.; da Silva, C.A.; Oba, E.T.; Pamplona, J.H.; Mela, M.; Doria, H.B.; Guiloski, I.C.; Ramsdorf, W.; Cestari, M.M. Hematologic and hepatic responses of the freshwater fish *Hoplias malabaricus* after saxitoxin exposure. *Toxicon* **2013**, *66*, 25–30. [CrossRef]

32. Maulvault, A.L.; Barbosa, V.; Alves, R.; Custódio, A.; Anacleto, P.; Repolho, T.; Pousão-Ferreira, P.; Rosa, R.; Marques, A.; Diniz, M. Ecophysiological responses of juvenile seabass (*Dicentrarchus labrax*) exposed to increased temperature and dietary methylmercury. *Sci. Total Environ.* **2017**, *586*, 551–558. [CrossRef]

33. Maulvault, A.L.; Barbosa, V.; Alves, R.; Anacleto, P.; Camacho, C.; Cunha, S.C.; Fernades, J.O.; Pousão-Ferreira, P.; Rosa, R.; Marques, A.; et al. Integrated multi-biomarker responses of juvenile seabass to diclofenac, warming and acidification co-exposure. *Aquat. Toxicol.* **2017**, *202*, 65–79. [CrossRef]

34. Pandey, S.; Ahmad, I.; Parvez, S.; Bin-Hafeez, B.; Haque, R.; Raisuddin, S. Effect of Endosulfan on Antioxidants of Freshwater Fish *Channa punctatus Bloch*: 1. Protection Against Lipid Peroxidation in Liver by Copper Preexposure. *Arch. Environ. Contam. Toxicol.* **2001**, *41*, 345–352. [CrossRef] [PubMed]

35. Silva, C.A.; Oba, E.T.; Ramsdorf, W.A.; Magalhães, V.F.; Cestari, M.M.; Ribeiro, C.A.O.; Silva de Assis, H.C. First report about saxitoxins in freshwater fish *Hoplias malabaricus* through trophic exposure. *Toxicon* **2011**, *57*, 141–147. [CrossRef] [PubMed]

36. Gubbins, M.J.; Eddy, F.B.; Gallacher, S.; Stagg, R.M. Paralytic shellfish poisoning toxins induce xenobiotic metabolising enzymes in Atlantic salmon (*Salmo salar*). *Mar. Environ. Res.* **2000**, *50*, 479–483. [CrossRef]

37. Madeira, D.; Vinagre, C.; Costa, P.M.; Diniz, M.S. Histopathological alterations, physiological limits, and molecular changes of juvenile *Sparus aurata* in response to thermal stress. *Mar. Ecol. Prog. Ser.* **2014**, *505*, 253–266. [CrossRef]

38. Jacob, D.; Petersen, J.; Eggert, B.; Alias, A.; Christensen, O.B.; Bouwer, L.M.; Braun, A.; Colette, A.; Déqué, M.; Georgievski, G.; et al. EURO-CORDEX: New high-resolution climate change projections for European impact research. *Reg. Environ. Chang.* **2014**, *14*, 563–578. [CrossRef]

39. Lawrence, J.F.; Niedzwiadek, B. Quantitative determination of paralytic shellfish poisoning toxins in shellfish by using prechromatographic oxidation and liquid chromatography with fluorescence detection. *J. AOAC Int.* **2001**, *84*, 1099–1108.

40. Bradford, M.M. A rapid and sensitive method for the quantification of microgram quantities of protein utilizing the principle of protein-dye binding. *Anal. Biochem.* **1976**, *72*, 248–254. [CrossRef]

toxins

Article

Discovery of a Potential Human Serum Biomarker for Chronic Seafood Toxin Exposure Using an SPR Biosensor

Kathi A. Lefebvre [1,*], Betsy Jean Yakes [2], Elizabeth Frame [3], Preston Kendrick [1], Sara Shum [4], Nina Isoherranen [4], Bridget E. Ferriss [1], Alison Robertson [5], Alicia Hendrix [6], David J. Marcinek [7] and Lynn Grattan [8]

[1] Environmental and Fisheries Sciences Division, Northwest Fisheries Science Center, National Marine Fisheries Service, National Oceanic and Atmospheric Administration, 2725 Montlake Blvd. East, Seattle, WA 98112, USA; kendrickps@gmail.com (P.K.); bridget.ferriss@noaa.gov (B.E.F.)
[2] U.S. Food and Drug Administration, Center for Food Safety and Applied Nutrition, College Park, MD 20740, USA; Betsy.Yakes@fda.hhs.gov
[3] Aquatic Toxicology Unit, King County Environmental Laboratory, Seattle, WA 98119, USA; Elizabeth.Frame@kingcounty.gov
[4] Department of Pharmaceutics, University of Washington, Seattle, WA 98195, USA; hms520@uw.edu (S.S.); ni2@uw.edu (N.I.)
[5] Department of Marine Sciences, University of South Alabama and the Dauphin Island Sea Lab, Dauphin Island, AL 36528, USA; arobertson@disl.org
[6] Department of Environmental and Occupational Health Sciences, University of Washington, Seattle, WA 98105-6099, USA; aliciah1@uw.edu
[7] Departments of Radiology and Bioengineering and Pathology, University of Washington Medical School, 850 Republican Street, Seattle, WA 98109, USA; dmarc@uw.edu
[8] Neurology Department, University of Maryland School of Medicine, Baltimore, MD 21201, USA; LGrattan@som.umaryland.edu
* Correspondence: Kathi.Lefebvre@noaa.gov

Received: 24 April 2019; Accepted: 21 May 2019; Published: 23 May 2019

Abstract: Domoic acid (DA)-producing harmful algal blooms (HABs) have been present at unprecedented geographic extent and duration in recent years causing an increase in contamination of seafood by this common environmental neurotoxin. The toxin is responsible for the neurotoxic illness, amnesic shellfish poisoning (ASP), that is characterized by gastro-intestinal distress, seizures, memory loss, and death. Established seafood safety regulatory limits of 20 μg DA/g shellfish have been relatively successful at protecting human seafood consumers from short-term high-level exposures and episodes of acute ASP. Significant concerns, however, remain regarding the potential impact of repetitive low-level or chronic DA exposure for which there are no protections. Here, we report the novel discovery of a DA-specific antibody in the serum of chronically-exposed tribal shellfish harvesters from a region where DA is commonly detected at low levels in razor clams year-round. The toxin was also detected in tribal shellfish consumers' urine samples confirming systemic DA exposure via consumption of legally-harvested razor clams. The presence of a DA-specific antibody in the serum of human shellfish consumers confirms long-term chronic DA exposure and may be useful as a diagnostic biomarker in a clinical setting. Adverse effects of chronic low-level DA exposure have been previously documented in laboratory animal studies and tribal razor clam consumers, underscoring the potential clinical impact of such a diagnostic biomarker for protecting human health. The discovery of this type of antibody response to chronic DA exposure has broader implications for other environmental neurotoxins of concern.

Keywords: chronic exposure; environmental neurotoxin; serum biomarker; seafood toxin; algal toxin; marine biotoxin

Key Contribution: Domoic acid (DA) is a potent neurotoxin that is naturally produced during harmful algal blooms and accumulates in filter-feeding shellfish. There is significant concern regarding the health impacts of chronic DA exposure with long-term shellfish consumption in coastal regions where consumers are known to consume low levels of DA year-round. Here we report the discovery of a DA-specific antibody in serum of chronic shellfish consumers that could serve as a diagnostic biomarker for chronic DA exposure and possible underlying health impacts. The antibody was detected in chronic shellfish consumers using a novel surface plasmon resonance (SPR) biosensor protocol. This discovery has broader diagnostic implications if the phenomenon also occurs with exposure to other environmental neurotoxins of concern.

1. Introduction

Domoic acid (DA) is a neurotoxin that is naturally produced during harmful algal blooms (HABs) by toxigenic *Pseudo-nitzschia* species with global distributions [1]. During HABs the toxin is transferred through food webs via filter-feeding pelagic and benthic species of finfish, shellfish, and other invertebrates to marine mammals, seabirds, and humans causing severe neurotoxicity and mortality [2–5]. Domoic acid poisoning in humans has been termed amnesic shellfish poisoning (ASP) and is characterized by gastrointestinal distress, seizures, permanent anterograde memory loss, death, and a host of other permanent neurological symptoms [6]. The first documented ASP event occurred in 1987 when over 100 people became ill and 4 died after consuming DA-contaminated mussels [4,7]. Follow up analyses of toxin levels in meal remnants, as related to human symptomology and additional non-human primate laboratory exposure studies, were performed to set a general seafood safety regulatory limit of 20 µg DA/g seafood, established by Health Canada in the late 1980s and rapidly adopted by the US Food and Drug Administration [8]. The regulatory limit of 20 µg DA/g of edible shellfish tissue was designed to protect seafood consumers from DA doses that would cause visible symptoms after a single average shellfish meal. It does not consider the potential health effects of repetitive or chronic long-term exposure to lower toxin concentrations [9,10]. This raises concerns regarding the potential health risks of chronic DA exposure to putatively "safe" concentrations of DA that are not currently being considered in regulations.

Recent findings in laboratory models as well as seafood consumption studies in humans have called attention to the importance of considering chronic low-level DA exposure in the management of health risks. In controlled laboratory studies, subclinical neurologic effects have been reported after long term low-level exposure to DA at doses below those that elicit the obvious clinical signs of ASP. For example, repetitive low-level exposure caused learning deficits and hyperactivity in adult mice [11] as well as neurobehavioral changes in neonatal mice with low-level exposures in utero [12]. Studies in a nonhuman primate model point to additional effects of chronic low-level exposure. Female *Macaca fascicularis* monkeys were given daily oral doses of DA near the current allowable daily intake regulatory level for humans through pregnancy and gestation. Cognitive assessments in the offspring revealed significant consequences on emerging memory processes in neonates [13]. In addition, structural and chemical changes in brain morphology and the development of intentional tremors were observed over time in the adult females that were exposed during pregnancy [13,14]. The potential effects of chronic DA exposure may extend to humans as well. Consumption studies in coastal dwelling human seafood consumers in the Pacific Northwest, USA, have documented chronic DA exposure via razor clams (a shellfish species known to retain low levels of DA up to a year after a HAB in this region) in both recreational and tribal harvesters [10,15], as well as an association between chronic DA exposure via razor clam consumption and memory decline [16,17].

The urgency for assessing chronic low-level DA exposure and associated health risks is further underscored by the increase in geographic extent, duration, and severity of HABs. Warmer ocean conditions have been linked to more frequent, longer lasting, geographically larger, and more toxic

Pseudo-nitzschia blooms, thereby increasing the risks of more frequent DA contamination of seafood resources [18–20]. Dietary exposures to the neurotoxin DA are inevitable for people consuming seafood harvested from regions where toxigenic *Pseudo-nitzschia* species occur, even with current regulatory limits enforced [10]. In order to effectively protect human health, both short-term recent exposure and long-term chronic exposure need to be considered when determining health risks and developing regulatory guidelines for safe seafood consumption. The subclinical effects of chronic low-level exposure to environmental neurotoxins are difficult to quantify in naturally exposed populations due to the multitude of confounding factors such as baseline health, alcohol and drug use, other contaminant exposures, age, and other general lifestyle choices. Unlike laboratory studies where exposure doses are controlled and defined, previous examinations of naturally exposed human populations have only been able to estimate DA exposure dose based on self-reported consumption rates and associated toxin levels in seafood reported by regulatory agencies. There is a critical need to develop tests for specific biomarkers that can verify chronic DA exposure in seafood consuming individuals.

In the present study, we collected two bodily fluids (blood and urine) in an effort to identify potential biomarkers of DA exposure in coastal dwelling Native Americans in Washington State who are at a particularly high risk of dietary DA exposure due to dependence on razor clams (RCs) as a food source. Razor clams in this region are known to contain low-levels of DA year-round [21]. In addition to bodily fluid samples, participants were given shellfish assessment surveys (SASs) to record recent RC consumption (within the last week) and long-term RC consumption (average monthly consumption over the last one to ten years). To test for a biomarker of chronic long-term DA exposure, human serum samples were evaluated for the presence of a DA-specific antibody. Additionally, human urine samples were tested for the presence of DA, which is rapidly eliminated via urine, to determine if DA is detectable after recent consumption and to confirm systemic DA exposure via consumption of legally-harvested shellfish.

2. Results

2.1. The DA Antibody Biomarker Was Detected in Some Chronic Shellfish Consumers via ELISA

Absorbance ratios indicative of DA-specific antibody presence were detected by enzyme-linked immunosorbent assay (ELISA) in three out of 42 serum samples collected in December 2015, four out of 40 samples collected in April 2016, and three out of 21 samples collected in May 2016. None of the serum samples collected in November 2012 ($n = 29$) were positive for antibody presence by ELISA. Absorbance ratios were calculated by dividing the RC consumer sample absorbance (X) by the sum of the mean non-RC consumer control serum absorbance and three times the standard deviation (X/0.153). Absorbance ratios greater than one in RC consumer serum samples indicated the presence of DA-specific antibodies. Control serum samples from non-RC consumers ($n = 31$) yielded a mean absorbance value of 0.078 ± 0.025 (SD). An additional blank ($n = 6$ over all assays), consisting of sample buffer only with no serum, yielded a mean absorbance value of 0.058 ± 0.002 (SD), suggesting that some non-specific binding occurs with control serum compared to buffer-only blanks. The small number of DA-specific antibody positive serum samples observed via ELISA revealed that a more sensitive method was needed for antibody detection. Consequently, a subset of samples was further analyzed via a surface plasmon resonance (SPR) biosensor.

2.2. The DA Antibody Biomarker Was Detected in a Majority of Chronic Shellfish Consumers via an SPR Biosensor

After the development of a more sensitive antibody detection method using a surface plasmon resonance (SPR) biosensor, a total of 61 serum samples from 22 individual RC consumers were analyzed via the SPR biosensor without the analyst's knowledge of the participants' consumption levels. All 22 individuals had serum drawn at a minimum of two and up to four possible timepoints (November 2012, December 2015, April 2016, and May 2016). After SPR analyses, long-term consumption was quantified for the 22 individuals. Twelve were categorized as high consumers (average consumption > 9 RCs per month over one to ten years), and ten were categorized as moderate consumers (average consumption of 3 to 9 RCs per month over one to ten years). Data from two of the high consumers were excluded from further analyses due to high nonspecific binding signals that prevented a determination of antibody presence or absence at multiple timepoints, and these results are not included in Table 1. Domoic acid-specific antibodies were detected in serum from at least one blood draw in 80% of the ten remaining high RC consumers and 40% of the moderate RC consumers tested (Table 1). In high consumers, 50% had the antibody present at all timepoints tested compared to 20% for moderate consumers (Table 1). Domoic acid-specific antibody presence was not detected in control samples ($n = 17$). Of the ten serum samples that tested positive by ELISA as described in Section 2.1., six of those also tested positive for antibody via SPR (Table 1). The other four ELISA-positive samples were not quantifiable via SPR due to high nonspecific binding (Table 1; participant ID 3 and another participant not included in Table 1 due to high non-specific binding at all timepoints). It is important to note that these reported percentages for antibody presence in chronic consumers are not representative of the population level prevalence due to limitations on the number of participants available for multiple blood draws, but these results do provide solid evidence for the development of the DA-specific antibody as a diagnostic biomarker.

Table 1. Presence of a domoic acid (DA)-specific antibody via a surface plasmon resonance (SPR) biosensor in multiple serum samples from 20 razor clam (RC) consumers. Consumers were classified as high or moderate consumers based on average monthly consumption rates recorded in yearly shellfish assessment surveys (SASs) over one to ten years. High = greater than nine RCs per month year-round, and moderate = three to nine RCs per month year-round.

Sample ID	Average Monthly Consumption	Antibody Presence (Nov 2012)	Antibody Presence (Dec 2015)	Antibody Presence (April 2016)	Antibody Presence (May 2016)	Years of SASs
1	High	Yes	Yes	Yes	Yes	9
2	High	Yes	Yes	Yes	ns	10
3	High	Yes	No	**+	**+	6
4	High	Yes	No	ns	ns	10
5	High	ns	Yes	Yes	Yes	2
6	High	ns	Yes+	Yes+	Yes+	1
7	High	ns	Yes+	Yes	ns	3
8	High	ns	No	Yes	No	2
9	High	ns	No	No	No	9
10	High	ns	No	No	ns	1
11	Moderate	Yes	Yes	ns	Yes+	8
12	Moderate	Yes	Yes	Yes+	ns	8
13	Moderate	Yes	Yes	No	ns	10
14	Moderate	Yes	No	ns	ns	6
15	Moderate	No	No	No	ns	7
16	Moderate	ns	No	No	No	1
17	Moderate	ns	No	No	No	1
18	Moderate	ns	No	ns	No	9
19	Moderate	ns	No	No	ns	2
20	Moderate	ns	No	No	ns	10

** = unable to determine due to high nonspecific binding; **YES** = tested positive for DA-specific antibody via an SPR biosensor; + = also tested positive for DA-specific antibody via ELISA; No = DA-specific antibody was not detected; ns = no sample available.

The SPR biosensor was a better method than ELISA for detecting DA-specific antibodies due to the increased specificity, ability to use smaller sample volumes, and incorporation of multiple tests per sample. When the serum antibody binds to the DA on the SPR chip surface, so can nonspecific interactants. This can make it difficult to determine if binding is due to a specific surface-bound DA and antibody interaction. The use of a secondary antibody enabled both signal enhancement and signal interference reduction, as the anti-human secondary antibody only binds to the human antibody from serum and not to the nonspecific binders (e.g., non-immunoglobulin serum proteins). The use of multiple tests per sample with three requirements for a positive result for DA-specific antibodies improved confidence and detection sensitivity.

In order to determine if a sample contained DA-specific antibodies, three questions were evaluated: (1) is there something that binds to the DA chip, (2) is that binding specific or nonspecific, and (3) is the binding to DA versus the chip? Two assays were performed for each sample in order to fully answer these questions. The first assay was a direct evaluation of antibody binding from serum. In order to be categorized as having antibody, the sample must show a binding response higher than that of the controls (direct binding; answering question 1) and also have the secondary antibody binding response higher than the controls (answering question 2; Figure 1). Both of these were quantified via the change in refractive index upon binding that is greater than control refractive index changes. The second assay used a high concentration of DA (10,000 ng/mL) pre-mixed with the serum sample in an inhibition assay format to determine if the serum antibody showed solution specificity to DA. A decrease in binding of the DA-mixed versus the non-DA mixed samples indicated that the binding seen on the SPR sensor surface was for DA and not something else on the sensor surface (answering question 3). When all three requirements were met, the sample was determined to contain DA antibody (Figure 1).

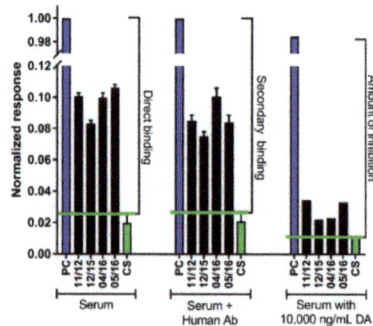

Figure 1. Example graph showing positive detection of a domoic acid (DA)-specific antibody (Ab) by surface plasmon resonance (SPR) in serum collected at four timepoints from a "high" chronic razor clam (RC) consumer (High = greater than nine RCs per month year-round; ID 1 in Table 1). The blue bars represent the mouse DA Ab positive control samples (PC). Black bars represent human seafood consumer serum at four collection dates (November 2012, December 2015, April 2016, and May 2016). The green bars represent controls (CS; human serum from seafood consumers from a region without DA blooms). Three analyses are shown: (1) direct binding, (2) secondary Ab binding, and (3) amount of inhibition. Horizontal green lines denote cutoff values based on control serum for direct binding (response at the end of the serum injection minus the starting baseline), secondary antibody binding (response at the end of the secondary antibody injection minus the starting baseline), and amount of inhibition (response difference between binding for the 0 ng/mL versus the addition of 10,000 ng/mL DA in serum samples). Error bars represent the standard deviation from triplicate measurement, and the data are normalized to the mouse DA-specific Ab control responses.

2.3. Recent RC Consumption and Detection of DA in Urine Confirms Systemic Exposure

Domoic acid was detected in urine samples from participants reported to have recently consumed RCs, confirming that systemic DA exposure occurs with consumption of RCs containing DA below the regulatory threshold of 20 µg DA/g shellfish (Figure 2). The percentage of people who had consumed RCs during the recent consumption target period (last nine days for December 2015 participants, and last seven days for April and May 2016 participants) were 39%, 40%, and 44% in December 2015 (*n* = 123 total), April 2016 (*n* = 69 total,) and May 2016 (*n* = 18 total), respectively. The percentages of people reporting recent RC consumption and with DA positive urine samples were 21%, 50%, and 25% for December 2015, April 2016, and May 2016, respectively (Figure 2). All urine samples that tested positive for DA via ELISA and high performance liquid chromatography/mass spectrometry (HPLC-MS/MS) (*n* = 30) were significantly positively correlated (Figure 3). The lower limits of quantitation (LLOQ) for urine samples were 0.4 ng/mL for ELISA and 0.3 ng/mL for HPLC-MS/MS. Figure 4 depicts chromatograms of the three DA transitions in spiked blank human urine and in human urine from RC consumers testing positive for DA. A subset of 16 urine samples that tested positive for DA above 1 ng DA/mL by ELISA and should have been detectable by HPLC-MS/MS, were negative by HPLC-MS/MS, suggesting that false positives can occur via ELISA methods and that HPLC-MS/MS is required for validation.

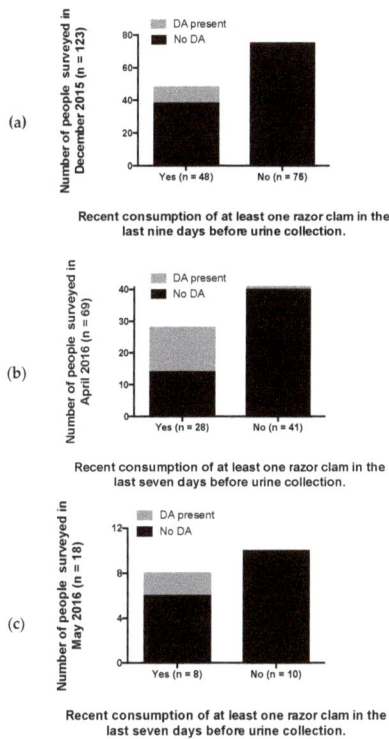

Figure 2. Bar graphs showing the number of people that consumed clams during the seven to nine days prior to urine sampling (Yes category) and the number of domoic acid (DA)-positive urine samples (gray portion in bar) collected in (**a**) December 2015, (**b**) April 2016, and (**c**) May 2016.

Total recent DA exposure for each individual was calculated from the SAS based on the number of RCs consumed (converted to total number of grams using 45 g edible meat per clam [9]) and the average DA concentrations reported in RCs for corresponding time periods and tribal harvest beaches by the Washington Department of Health (WDOH) biotoxin monitoring program (23 ppm in December 2015, 6 ppm in April 2016, and 10 ppm in May 2016). Maximum DA exposures were 35, 11, and 6.3 mg of DA per person for target periods in December 2015, April 2016, and May 2016, respectively. Total milligrams of DA consumed per person were positively correlated with DA concentrations detected in corresponding urine samples in December 2015 and April 2016 (Figure 4). Domoic acid exposure and urine DA concentrations were not significantly correlated in the May samples due to a small sample size ($n = 18$) and a smaller range of DA exposures compared to the December ($n = 123$) and April ($n = 69$) timepoints (Figure 5).

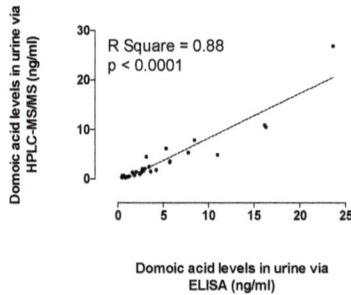

Figure 3. Comparison of domoic acid (DA) concentrations in urine samples ($n = 30$) quantified via Biosense DA enzyme-linked immunosorbent assays (ELISA) and high performance liquid chromatography tandem mass spectrometry (HPLC-MS/MS).

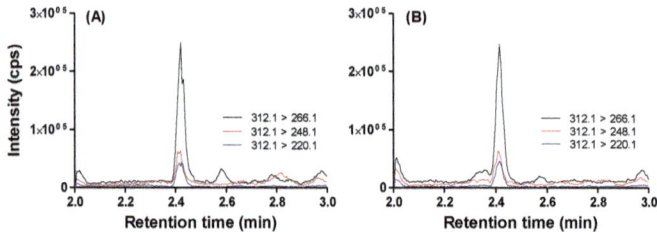

Figure 4. High performance liquid chromatography tandem mass spectrometry (HPLC-MS/MS) chromatograms of three domoic acid (DA) transitions (m/z 312.1 > 266.1, 248.1, 220.1) in (**A**) spiked urine with 19.9 ng DA/mL, and (**B**) human urine sample measured with 20.2 ng DA/mL.

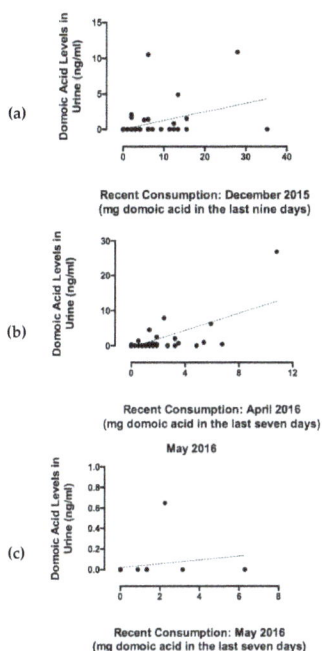

(a)

(b)

(c)

Figure 5. Comparison of total domoic acid exposure in the previous nine days in people surveyed in (**a**) December 2015, or previous seven days in people surveyed in (**b**) April and (**c**) May 2016, and toxin levels quantified via HPLC-MS/MS in corresponding urine samples collected on the same day that participants completed the recent consumption surveys.

3. Discussion

This study is the first to provide evidence for a DA-specific antibody response to chronic DA exposure in humans and, to our knowledge, is the first to detect and quantify the toxin in urine of naturally-exposed human seafood consumers. The detection of DA in urine and a DA-specific antibody in serum in multiple human shellfish consumers unequivocally confirms that Pacific Northwest coastal RC harvesters are systemically-exposed to DA via consumption of shellfish that contain toxin concentrations below the seafood safety regulatory limit and therefore deemed "safe" to consume. Moreover, these measures may serve as important biomarkers for diagnosing both recent and long-term chronic exposure, respectively (Figure 2; Table 1). Increased understanding of the exposure rates and the health impacts of both acute, high-level and chronic, low-level DA exposure are critical for effectively managing health risks to seafood consumers.

3.1. Detection of a DA-Specific Antibody in Serum Indicates Chronic Exposure

The detection of a DA-specific antibody in serum from multiple long-term RC consumers reveals a promising diagnostic tool for identifying chronically-exposed people. The first evidence for an immune response and DA-specific antibody production with chronic DA exposure was found during controlled laboratory studies by our research team using the zebrafish (*Danio rerio*) model system. Zebrafish were exposed once a week for multiple months to low doses of DA that were below doses that induce visible signs of neurotoxicity. Chronic DA exposure to these subclinical doses induced a significant immune response as indicated at the transcriptional level by whole-genome microarray profiling that was temporally linked to evidence of DA-specific antibody presence in serum [22]. Immune function genes were significantly upregulated after 18 weeks of exposure followed by evidence for the antibody in serum detected at the next sampling timepoint of 24 weeks of exposure, suggesting multiple weeks of low-level exposure were required for antibody production [22]. This discovery that long-term low-level DA exposure may lead to the induction of DA-specific antibody production in serum provided the impetus for exploring its use as a biomarker for chronic exposure in mammalian species using naturally-exposed California sea lions (*Zalophus californianus*). We found that 65% of sea lions known to have previous DA exposure tested positive for antibody presence while none of the non-exposed captive control sea lions tested positive, providing further validation of the potential biomarker [22]. These early studies in zebrafish and sea lions utilized ELISA techniques for antibody recognition and laid the groundwork for our investigations in human consumers. In the present study, we developed more sensitive and specific detection methods using an SPR biosensor that has provided compelling evidence for the antibody response in naturally exposed humans (Table 1).

The apparent transient nature of the DA-specific antibody presence in serum of chronic RC consumers is an attribute that makes it potentially more valuable as a clinical diagnostic biomarker (Table 1). All serum samples used for antibody detection were analyzed without knowledge of consumption status. We selected individuals for whom we had been able to collect serum at two or more timepoints in order to assess the consistency of antibody presence. After determination of antibody presence or absence in serum, long-term consumption data from the SASs were then used to determine chronic consumption status for each individual (Table 1). Fifty percent of high chronic consumers (those eating on average greater than 9 RCs per month year-round) and 20% of moderate RC consumers (those eating an average of 3 to 9 RCs per month year-round) tested positive for antibody presence at all timepoints. This is consistent with the assumption that greater exposure would be linked with increased prevalence of the antibody. In both high and moderate consumer categories, some individuals did not have detectable DA-specific antibody presence at any timepoint, while some were positive at some timepoints and negative at others (Table 1). These findings are consistent with the idea that antibody presence (and likely concentration) will vary in a predictable relationship with chronic exposure levels and duration. This is a critical element for establishing a biomarker of exposure. While we do not have the statistical power in this study to determine this relationship, our data provide strong evidence for the value of future studies with a larger population size.

3.2. Uses for a Biomarker of Chronic DA Exposure

The proposed DA-specific antibody biomarker would be valuable for identifying chronic exposure risks and, potentially, as a diagnostic indicator of underlying toxicological insult. Previous work in laboratory mouse models, using chronic low-level exposure paradigms, have documented novel effects of chronic exposure to DA at doses below those that induce visible outward signs of toxicity that are characteristic of ASP. These neurobehavioral effects of chronic low-level DA exposure included significant spatial learning and memory deficits as well as hyperactivity [11]. Unlike the permanent neurologic damage and hippocampal lesions documented with high-level seizure-inducing DA exposure [23–25], both hyperactivity and cognitive deficits observed with chronic low-level exposure were found to be reversible after a recovery period of no exposure and there were no gross morphologic or neuroinflammatory alterations in the hippocampal regions [11,26]. These reversible neurobehavioral effects were associated with a selective increase in vesicular glutamate transporter (VGluT1) levels within VGluT1-expressing boutons in the CA1 region of the hippocampus following repeated, low-level DA exposure in this model [26]. This could be responsible for creating a more excitatory environment in the brain through increased glutamate release from presynaptic excitatory boutons, thereby contributing to the observed hyperactivity and cognitive deficits observed [26]. If these patterns hold true for human seafood consumers, and if the DA-specific antibody is elevated in serum when cognitive deficits are present, but absent or reduced after recovery, then the biomarker would be a valuable diagnostic tool for chronic disease and subsequent recovery.

3.3. Detection of DA in Urine Confirms Systemic Exposure and Recent Consumption

The detection of DA in urine confirms that systemic DA exposure occurs via consumption of shellfish containing DA concentrations below current regulatory thresholds (20 µg DA/g) and may be useful for indicating recent exposure if sampled soon after consumption. The high variability of DA presence and level in RC consumer urine suggests that DA levels in urine are not accurate indicators of recent exposure over a seven-to-nine day period (Figures 2 and 5). However, the fact that DA was detectable with consumption of legally-harvested seafood, suggests that detection of DA in urine may be useful as a shorter-term marker (i.e., within 24 h of consumption) for acute exposure concerns. Several factors confounded the relationship between DA concentrations in urine and DA consumption as quantified in the present study, such as toxin excretion rates, time after last meal, and rates of urination between exposure and sample collection. Domoic acid is rapidly excreted in the urine of mammals with a majority of the toxin eliminated within 24 h [27]. Rapid depuration rates along with the other listed factors account for the high variability in the presence and level of toxin detected in urine. The original goal was to obtain enough data to make comparisons of DA concentrations in urine and DA consumption within the last 24 h. However, the participant group taking the recent shellfish consumption surveys did not contain enough individuals with consumption within the last 24 h, thus requiring us to consolidate consumption over the week-long period. Even with a longer time period, DA was detectable in urine and correlated to consumption rates (Figure 5). We expect that urine sampled within 24 h of exposure would provide more consistent results, suggesting that DA in urine may be valuable as a diagnostic for acute exposure if urine is collected immediately after a patient feels ill from consuming shellfish. The use of urine DA tests would be a valuable component to assessing potential cases of DA neuroexcitotoxicity if collected within a short time window after suspected exposure. Indeed, domoic acid concentrations in urine, stomach contents, and feces have been used in assessments of marine mammal health for years and are part of a diagnostic protocol along with several other metrics for identifying acute DA poisoning in California sea lions [28,29].

4. Conclusions

In summary, our discovery of a DA-specific antibody in the serum of shellfish consumers is a breakthrough that could be used for the development of a diagnostic tool for assessing chronic DA exposure risks for which there are currently no protections. Our novel detection of DA in urine of naturally-exposed human shellfish consumers confirms that systemic DA exposure occurs via consumption of legally-harvested shellfish and represents a chronic exposure risk. The presence of a DA-specific antibody in serum provides an additional tool for studies to assess the relationship between chronic DA exposure and cognitive function in naturally exposed human populations. Future studies will also address whether the reduction or absence of the DA antibody is indicative of recovery following chronic DA exposure.

5. Materials and Methods

5.1. Quantification of Long-Term RC Consumption and Chronic Exposure

Shellfish assessment surveys (SAS) [30] to quantify long term razor clam (RC) consumption were administered to coastal dwelling Native Americans in Washington State as part of the "communities advancing the studies of tribal nations across their lifespan" (CoASTAL) study [31]. The CoASTAL participants completed yearly SASs, recording monthly RC consumption for as many as ten consecutive years. Average monthly RC consumption rates over one to ten years were then used to identify high consumers (those eating > nine RCs per month year-round) or moderate consumers (those eating three to nine RCs per month year-round).

5.2. Blood Collection for Biomarkers of Long-Term Chronic DA Exposure

Serum samples were collected from individuals at four sampling periods in conjunction with the CoASTAL study: in November 2012, December 2015, April 2016, and May 2016. Blood samples were collected by certified phlebotomists using BD Vacutainer Blood Collection kits into BD Vacutainer serum collection tubes. Whole blood was allowed to sit undisturbed at room temperature for 15–30 min to allow for clotting, followed by centrifugation at 4000× g for 15 min in a refrigerated centrifuge to remove the clot. Serum was immediately transferred and aliquoted into clean cryovials and frozen at −20 °C until further analyses via ELISA and/or SPR biosensor methods. An additional set of serum samples (n = 31) collected from moderate seafood consumers in Florida, where DA is not commonly observed, were opportunistically obtained and used as controls. Control blood samples were collected from volunteers with BD Vacutainer Safety-Lok blood collection kits into BD Vacutainer K2 EDTA spray coated tubes. Samples were mixed by inversion 8–10 times immediately after collection to ensure protein stabilization and anticoagulant action and then centrifuged at 2000× g for 15 min at 4 °C to separate serum and deplete platelets. Serum was aseptically removed and stored as aliquots at −20 °C. Before all analyses, serum samples were purified with a NAb Protein G Spin Kit per the vendor protocols (Thermo Fisher Scientific, Waltham, MA, USA) in order to remove extraneous proteins that may interfere with antibody binding and to enable an appropriate buffer for analysis. Cleaned-up serum was eluted in the standard amine-based elution buffer (pH 2.8) that was then neutralized with 10% (v/v) 1 M Tris HCl, pH 8.5 buffer.

5.3. Domoic Acid-Specific Antibody Presence via ELISA

Serum samples from RC consumers from Washington State collected in November 2012 (n = 29), December 2015 (n = 42), April 2016 (n = 40), and May 2016 (n = 21), as well as control serum samples (n = 31) collected from seafood consumers from Florida were analyzed for the presence of a DA-specific antibody via an enzyme-linked immunosorbent assay (ELISA) format. Detection specificity of DA-specific human IgG was accomplished using a DA conjugated 96-well plate obtained from the commercially available ASP ELISA kit for quantitative determination of domoic acid from Biosense Laboratories. Serum was applied to sample wells and DA-specific antibodies were allowed to bind to the DA conjugated to the plate. After thorough washing, HRP labeled anti-human IgG (goat anti human IgG H + L (HRP) from abcam, ab97161) diluted 1:5000 in 1% ovalbumin in PBS-T was allowed to incubate for one hour at room temperature, then TMB substrate was added, and absorbance was detected with a BioTek Epoch Plate reader. Absorbance ratios greater than one were used to determine antibody presence and were calculated by dividing RC consumer serum sample absorbance by the mean control absorbance plus three times the standard deviation from 31 control serum samples.

5.4. Domoic Acid-Specific Antibody Presence via an SPR Biosensor

A subset of control serum samples (n = 17) and serum samples from RC consumers (n = 22) in which samples were taken at more than one timepoint for each individual were analyzed for the presence of a DA-specific antibody via a surface plasmon resonance (SPR) biosensor (T200, GE Healthcare, Pittsburgh, PA, USA). Samples were run in random order during three consecutive days using a previously developed SPR biosensor surface [32] and a modified assay for detecting DA antibody in the complex serum matrix. The sensor surface had three DA-conjugated channels and one reference channel. Each sample was injected over all four flow cells, secondary antibody was then introduced, and then regeneration solution pulled both off the complex leaving the DA surface ready for the next sample.

Two vials of cleaned-up serum were diluted 1:5 in HBS-EP+ (GE Healthcare, pH 7.4) running buffer. Two aliquots of samples were used: the first aliquot vial was mixed with 10% (v/v) with 0 ng DA/mL, while the second was mixed with 10,000 ng DA/mL (Sigma-Aldrich, St. Louis, MO, USA; DA diluted in HBS-EP+) just prior to analysis. This allowed for two analyses of each sample with the first being a direct binding assay and the second being an inhibition assay evaluation. As shown in Figure 6, a sample cycle consisted of a binding time of 120 s (flow rate of 25 µL/min), followed by a 90 s binding of secondary antibody (30 µg/mL goat anti-human IgG H&L (abcam, ab97161), flow rate of 10 µL/min), and then regeneration for 60 s with 50 mM HCl, 0.5% SDS (flow rate of 25 µL/min). The secondary antibody was employed to increase the signal intensity, while simultaneously lowering background interference from non-specific binding. Positive control samples of mouse anti-DA (1:800 dilution in HBS-EP+; information on how this antibody was produced and evaluated is found in reference 32) with 30 µg/mL secondary, rabbit anti-mouse (GE Healthcare, mouse antibody capture kit, BR-1008-38), and negative control samples (buffer and non-exposed human serum) were interspersed with the RC consumer and control samples to ensure assay functionality and chip stability.

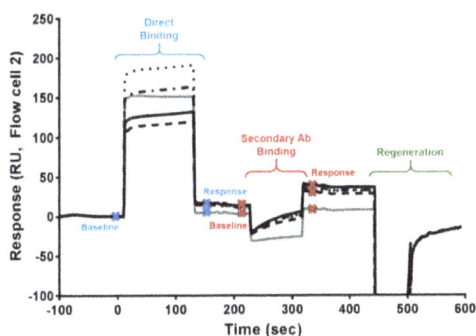

Figure 6. Overlay plot of sensorgrams showing the analysis cycle: direct binding, secondary antibody binding, and regeneration. Respective baselines and responses (blue = direct binding and red = secondary antibody binding) used to calculate the SPR signal are marked with "X" on the curves. The black lines (dotted, dash-dot, solid, and dashed) are representative sensorgrams from the four serum samples collected from a "high" RC consumer (ID 1 in Table 1 and Figure 1): November 2012 sample (dotted curve); December 2015 sample (dashed curve), April 2016 sample (solid curve), and May 2016 sample (dashed-dotted curve). The grey curve illustrates a representative sensorgram for a control serum sample (CS; human serum from a seafood consumer from a region without DA blooms; same sample as displayed in Figure 1). A simple visual evaluation illustrates that the "high" RC consumer, at all timepoints, had more SPR signal at both the direct binding response (blue Xs) and the secondary antibody binding response (red Xs) steps than the CS sample. These values are quantified and graphed in Figure 1.

To evaluate the data, control sample averages and standard deviations were obtained to create cutoff values for direct binding (response at end of the serum injection minus the starting baseline), secondary antibody binding (response at the end of the secondary antibody injection minus the starting baseline), and amount of inhibition (response difference between binding for the 0 ng/mL versus the 10,000 ng/mL DA in serum samples; Figure 1). Controls run on Days 1 and 2 ($n = 8$) were used to determine an average cutoff value for the initial binding, secondary antibody, and inhibition responses. As the chip showed minor degradation in binding in the positive control, mouse-DA antibody on Day 3, separate control samples ($n = 11$) analyzed on Day 3 were used to create the corresponding cutoff values for data generated on Day 3.

5.5. Quantification of Recent RC Consumption and DA Exposure

Shellfish assessment surveys, designed to quantify recent RC consumption, were also completed by CoASTAL participants at three sampling dates. In order to quantify recent DA exposure, the number of RCs consumed, as well as the source beaches for harvesting, were recorded for target periods of the last nine days for December 2015 surveys and the last seven days for April and May

2016 surveys. Recent DA exposure was calculated for each individual using the total number of grams of RCs consumed in the target period multiplied by the DA levels quantified in RCs from the source harvest beaches, as reported by the Washington State Department of Health (WDOH) Biotoxin Monitoring Program.

5.6. Quantification of DA in Urine as an Indicator of Systemic Exposure and Recent Consumption

Urine samples were collected in sterile 100 mL urine collection cups from individuals on the same day that they completed the recent RC consumption surveys (n = 123 in December 2015, n = 69 in April 2016, and n = 18 in May 2016). Immediately following collection, urine samples were cooled on ice and frozen with dry ice until placed in laboratory freezers at −20 °C. Domoic acid was quantified in urine samples using a commercially available enzyme-linked immunosorbent assay (ELISA) for DA from Biosense as per kit instructions [33] and validated via high performance liquid chromatography tandem mass spectrometry (HPLC-MS/MS). For ELISA methods, DA was extracted from urine via standard procedures using a 1:4 ratio of sample to 50% MeOH extraction solvent [34]. Final extracts were further diluted 10-fold in dilution buffer before quantification. For HPLC-MS/MS analyses, the samples were analyzed using a recently developed and validated HPLC-MS/MS method [35]. In brief, urine samples were extracted with 100% methanol at 1:1 v/v ratio. Samples were vortexed for 15 s and subsequently centrifuged at 16,100× g for 15 min. Supernatant was collected for analysis. Standard curves were prepared in naïve human urine with spiked domoic acid concentration ranging between 0.3 and 40 ng/mL. Samples were analyzed on Shimadzu UFLC XR DGU-20A5 (Shimadzu Scientific Instruments, Pleasanton, CA, USA) equipped with Synergi Hydro-RP 100 Å LC column (2.5 μm, 50 mm × 2 mm; Phenomenex) with a guard cartridge (2 × 2.1 mm, sub 2 μm; Phenomenex, Torrance, CA, USA) connected in line with AB Sciex 6500 qTrap Q-LIT mass spectrometer (AB Sciex, Concord, Ontario, Canada). The HPLC method uses a 9 min gradient running at 0.5 mL/min. The gradient initiates at 95% (A) water with 0.1 formic acid and 5% (B) 95:5 (v/v) acetonitrile:water with 0.1% formic acid for a minute, gradually increases to 100% B over the next 3 min, continues at 100% B for 30 s before returning to 5% B over the next 30 s, and runs at initial condition for another 4 min. Domoic acid was ionized using electrospray ionization (ESI) operating in positive ion mode and was monitored using multiple reaction monitoring (MRM) for m/z transition 312.1 > 266.1, m/z 321.1 > 248.1, and 312.1 > 220.1 (Figure 4). The urine calibration standards and quality control samples were prepared by spiking blank urine with the authentic certified reference standards as described previously.

Author Contributions: All co-authors contributed substantially to this research and manuscript in the following ways; Conceptualization, K.A.L., B.J.Y. and L.G.; Methodology, K.A.L., B.J.Y., E.F., P.K., S.S., N.I. and D.J.M.; Validation, K.A.L., B.J.Y., S.S., N.I. and A.R.; Formal Analysis, K.A.L., B.J.Y., E.F., P.K., S.S., N.I., B.E.F., A.R., A.H. and L.G.; Investigation, K.A.L., B.J.Y., E.F., B.E.F., A.R., A.H., D.J.M. and L.G.; Resources, K.A.L., N.I., D.J.M. and L.G.; Data Curation, K.A.L., B.J.Y., E.F., P.K., S.S., B.E.F. and L.G.; Writing-Original Draft Preparation, K.A.L., B.J.Y., S.S. and A.R.; Writing-Review & Editing, K.A.L., B.J.Y., E.F., S.S., N.I., A.R., D.J.M. and L.G.; Supervision, K.A.L., N.I., D.J.M. and L.G.; Project Administration, K.A.L. and L.G.; Funding Acquisition, K.A.L., N.I., D.J.M. and L.G.

Funding: This research was funded by the National Institutes of Health (NIH) and the National Institutes of Environmental Health Sciences (NIEHS) 1R01ES012459 (to Lynn Grattan), R01ES023043 (to Nina Isoherranen and Sara Shum), the NIH R01s ES021930 and ES030319 (to David J. Marcinek and Kathi A. Lefebvre), and the National Science Foundation (NSF) OCE-1314088 and OCE-1839041 (to David J. Marcinek and Kathi A. Lefebvre). Control sample collection was made possible through a collaborative Food and Drug Administration (FDA) and Center for Disease Control (CDC) project 1 U0 EH000421-01 (to Lynn Grattan, Alison Robertson, and PI: J.G. Morris) under an approved institutional review board (IRB) from the University of Maryland HP 00043841 (to Lynn Grattan). Informed consent forms were completed by all participants in this study.

Acknowledgments: The authors would like to thank the Quinault Indian Nation for their invaluable contributions to this research without which this discovery would not have been possible. Special thanks to Joe Schumacker (Quinault Department of Fisheries), Dawn Radonski (Quinault Tribal Nation), and Laura Castellon and Sparkle Roberts (University of Maryland).

Conflicts of Interest: The authors declare no conflict of interest.

References

1. Trainer, V.L.; Hickey, B.M.; Bates, S.S. Toxic diatoms. In *Oceans and Human Health: Risks and Remedies from the Sea*; Walsh, P.J., Smith, S.L., Fleming, L.E., Solo-Gabriele, H., Gerwick, W.H., Eds.; Elsevier Science Publishers: New York, NY, USA, 2008; pp. 219–238.
2. Scholin, C.A.; Gulland, F.; Doucette, G.J.; Benson, S.; Busman, M.; Chavez, F.P.; Cordaro, J.; DeLong, R.; De Vogelaere, A.; Harvey, J.; et al. Mortality of sea lions along the central california coast linked to a toxic diatom bloom. *Nature* **2000**, *403*, 80–84. [CrossRef] [PubMed]
3. Work, T.M.; Barr, B.; Beale, A.M.; Fritz, L.; Quilliam, M.A.; Wright, J.L.C. Epidemiology of domoic acid poisoning in brown pelicans (*Pelicanus occidentalis*) and brandt's cormorants (*Phalacrocorax penicillatus*) in california. *J. Zoo Wild. Med.* **1993**, *24*, 54–62.
4. Perl, T.M.; Bedard, L.; Kosatsky, T.; Hockin, J.C.; Todd, E.C.; Remis, R.S. An outbreak of toxic encephalopathy caused by eating mussels contaminated with domoic acid. *N. Engl. J. Med.* **1990**, *322*, 1775–1780. [CrossRef]
5. Lefebvre, K.A.; Powell, C.L.; Busman, M.; Doucette, C.J.; Moeller, P.D.R.; Sliver, J.B.; Miller, P.E.; Hughes, M.P.; Singaram, S.; Silver, M.W.; et al. Detection of domoic acid in northern anchovies and california sea lions associated with an unusual mortality event. *Nat. Toxins* **1999**, *7*, 85–92. [CrossRef]
6. Todd, E.C.D. Domoic acid and amnesic shellfish poisoning: A review. *J. Food Prot.* **1993**, *56*, 69–83. [CrossRef] [PubMed]
7. Todd, E. Amnesic shellfish poisoning—A new seafood toxin syndrome. In *Toxic Marine Phytoplankton*; Graneli, E., Sundstrom, B., Edler, L., Anderson, D.M., Eds.; Elsevier: New York, NY, USA, 1990; pp. 504–508.
8. Wekell, J.C.; Hurst, J.; Lefebvre, K.A. The origin of the regulatory limits for PSP and ASP toxins in shellfish. *J. Shellfish Res.* **2004**, *23*, 927–930.
9. Marien, K. Establishing tolerable dungeness crab (Cancer magister) and razor clam (Siliqua patula) domoic acid contaminant levels. *Environ. Health Perspect.* **1996**, *104*, 1230–1236.
10. Ferriss, B.E.; Lefebvre, K.A.; Ayres, D.; Borchert, J.; Marcinek, D.J. Acute and chronic dietary exposure to domoic acid in recreational harvesters: A survey of shellfish consumption behavior. *Environ. Int.* **2017**, *101*, 70–79. [CrossRef]
11. Lefebvre, K.A.; Kendrick, P.S.; Ladiges, W.C.; Hiolski, E.M.; Ferriss, B.E.; Smith, D.R.; Marcinek, D.J. Chronic low-level exposure to the common seafood toxin domoic acid causes cognitive deficits in mice. *Harmful Algae* **2017**, *64*, 20–29. [CrossRef]
12. Shiotani, M.; Cole, T.B.; Hong, S.; Park, J.J.Y.; Griffith, W.C.; Burbacher, T.M.; Workman, T.; Costa, L.G.; Faustman, E.M. Neurobehavioral assessment of mice following repeated oral exposures to domoic acid during prenatal development. *Neurotoxicol. Teratol.* **2017**, *64*, 8–19. [CrossRef]
13. Burbacher, T.M.; Grant, K.S.; Petroff, R.; Shum, S.; Crouthamel, B.; Stanley, C.; McKain, N.; Jing, J.; Isoherranen, N. Effects of oral domoic acid exposure on maternal reproduction and infant birth characteristics in a preclinical nonhuman primate model. *Neurotoxicol. Teratol.* **2018**, *72*, 10–21. [CrossRef]
14. Petroff, R.; Richards, T.; Crouthamel, B.; McKain, N.; Stanley, C.; Grant, K.S.; Burbacher, T.M. Chronic, low-level oral exposure to marine toxin, domoic acid, alters whole brain morphometry in nonhuman primates. *Neurotoxicol. Teratol.* **2018**, *72*, 114–124. [CrossRef]
15. Boushey, C.J.; Delp, E.J.; Ahmad, Z.; Wang, Y.; Roberts, S.M.; Grattan, L.M. Dietary assessment of domoic acid exposure: What can be learned from traditional methods and new applications for a technology assisted device. *Harmful Algae* **2016**, *57*, 51–55. [CrossRef]
16. Grattan, L.M.; Boushey, C.; Tracy, K.; Trainer, V.L.; Roberts, S.M.; Schluterman, N.; Morris, J.G., Jr. The association between razor clam consumption and memory in the coastal cohort. *Harmful Algae* **2016**, *57*, 20–25. [CrossRef]
17. Grattan, L.; Boushey, C.; Liang, Y.; Lefebvre, K.; Castellon, L.; Roberts, K.; Toben, A.; Morris, J. Repeated dietary exposure to low levels of domoic acid and problems with everyday memory: Research to public health outreach. *Toxins* **2018**, *10*, 103. [CrossRef]
18. McCabe, R.M.; Hickey, B.M.; Kudela, R.M.; Lefebvre, K.A.; Adams, N.G.; Bill, B.D.; Gulland, F.D.M.; Thomson, R.E.; Cochlan, W.P.; Trainer, V.L. An unprecedented coastwide toxic algal bloom linked to anomalous ocean conditions. *Geophys. Res. Lett.* **2016**, *43*. [CrossRef]

19. Moore, S.K.; Trainer, V.L.; Mantua, N.J.; Parker, M.S.; Laws, E.A.; Backer, L.C.; Fleming, L.E. Impacts of climate variability and future climate change on harmful algal blooms and human health. *Environ. Health* **2008**, *7* (Suppl. 2), S4. [CrossRef]

20. Van Dolah, F.M. Marine algal toxins: Origins, health effects, and their increased occurrence. *Environm. Health Perspect.* **2000**, *108*, 133–141. [CrossRef]

21. Wekell, J.C.; Gauglitz, E.J., Jr.; Barnett, H.J.; Hatfield, C.L.; Simons, D.; Ayres, D. Occurrence of domoic acid in washington state razor clams (*siliqua patula*) during 1991–1993. *Nat. Toxins* **1994**, *2*, 197–205. [CrossRef]

22. Lefebvre, K.A.; Frame, E.R.; Gulland, F.; Hansen, J.D.; Kendrick, P.S.; Beyer, R.P.; Bammler, T.K.; Farin, F.M.; Hiolski, E.M.; Smith, D.R.; et al. A novel antibody-based biomarker for chronic algal toxin exposure and sub-acute neurotoxicity. *PLoS ONE* **2012**, *7*, e36213. [CrossRef]

23. Silvagni, P.A.; Lowenstine, L.J.; Spraker, T.; Lipscomb, T.P.; Gulland, F. Pathology of domoic acid toxicity in california sea lions (*Zalophus californianus*). *Vet. Pathol.* **2005**, *42*, 184–191. [CrossRef]

24. Scallet, A.C.; Binienda, Z.; Caputo, F.A.; Hall, S.; Paule, M.G.; Rountree, R.L.; Schmued, L.; Sobotka, T.; Slikker, W., Jr. Domoic acid-treated cynomolgus monkeys (*M. Fascicularis*): Effects of dose on hippocampal neuronal and terminal degeneration. *Brain Res.* **1993**, *627*, 307–313. [CrossRef]

25. Strain, S.M.; Tasker, R.A. Hippocampal damage produced by systemic injections of domoic acid in mice. *Neuroscience* **1991**, *44*, 343–352. [CrossRef]

26. Moyer, C.E.; Hiolski, E.M.; Marcinek, D.J.; Lefebvre, K.A.; Smith, D.R. Repeated low level domoic acid exposure increases ca1 vglut1 levels, but not bouton density, vglut2 or vgat levels in the hippocampus of adult mice. *Harmful Algae.* in press. [CrossRef]

27. Suzuki, C.A.; Hierlihy, S.L. Renal clearance of domoic acid in the rat. *Food Chem. Toxicol.* **1993**, *31*, 701–706. [CrossRef]

28. Gulland, F.M.; Haulena, M.; Fauquier, D.; Langlois, G.; Lander, M.E.; Zabka, T.; Duerr, R. Domoic acid toxicity in californian sea lions (*Zalophus californianus*): Clinical signs, treatment and survival. *Vet. Rec.* **2002**, *150*, 475–480. [CrossRef]

29. Zabka, T.S.; Goldstein, T.; Cross, C.; Mueller, R.W.; Kreuder-Johnson, C.; Gill, S.; Gulland, F.M.D. Characterization of a degenerative cardiomyopathy associated with domoic acid toxicity in california sea lions (*Zalophus californianus*). *Vet. Pathol.* **2009**, *46*, 105–119. [CrossRef]

30. Fialkowski, M.K.; McCrory, M.A.; Roberts, S.M.; Tracy, J.K.; Grattan, L.M.; Boushey, C.J. Evaluation of dietary assessment tools used to assess the diet of adults participating in the communities advancing the studies of tribal nations across the lifespan cohort. *J. Am. Diet. Assoc.* **2010**, *110*, 65–73. [CrossRef]

31. Tracy, K.; Boushey, C.J.; Roberts, S.M.; Morris, J.G.; Grattan, L.M. Communities advancing the studies of tribal nations across their lifespan: Design, methods, and baseline of the coastal cohort. *Harmful Algae* **2016**, *57*, 9–19. [CrossRef]

32. Yakes, B.J.; Buijs, C.T.; Elliott, C.T.; Campbell, K. Surface plasmon resonance biosensing: Approaches for screening and characterising antibodies for food diagnostics. *Talanta* **2016**, *156*, 55–63. [CrossRef] [PubMed]

33. Lefebvre, K.A.; Robertson, A.; Frame, E.R.; Colegrove, C.M.; Nance, S.; Baugh, K.A.; Wiedenhoft, A.; Gulland, F.M.D. Clinical signs and histopathology associated with domoic acid poisoning in northern fur seals (*Callorhinus ursinus*) and comparison of toxin detection methods. *Harmful Algae* **2010**, *9*, 374–383. [CrossRef]

34. Wright, J.L.C.; Quilliam, M.A. Methods for domoic acid, the amnesic shellfish poisons. In *Manual on Harmful Marine Microalgae*; Hallegraeff, G., Anderson, D.M., Cembella, A.D., Eds.; Intergovernmental Oceanographic Commission Manuals and Guides No. 33; UNESCO: Paris, France, 1995; pp. 113–133.

35. Shum, S.; Kirkwood, J.S.; Jing, J.; Petroff, R.; Crouthamel, B.; Grant, K.S.; Burbacher, T.M.; Nelson, W.L.; Isoherranen, N. Validated hplc-ms/ms method to quantify low levels of domoic acid in plasma and urine after subacute exposure. *ACS Omega* **2018**, *3*, 12079–12088. [CrossRef]

toxins

MDPI

Article

Toxicity and Toxin Composition of the Greater Blue-Ringed Octopus *Hapalochlaena lunulata* from Ishigaki Island, Okinawa Prefecture, Japan

Manabu Asakawa [1,*], Takuya Matsumoto [2], Kohei Umezaki [1], Kyoichiro Kaneko [1], Ximiao Yu [1], Gloria Gomez-Delan [3], Satoshi Tomano [4], Tamao Noguchi [5] and Susumu Ohtsuka [6]

[1] Laboratory of Marine Bioresource Chemistry, Graduate School of Biosphere Science, Hiroshima University, Higashi-Hiroshima 739-8528, Japan; m186510@hiroshima-u.ac.jp (K.U.); m181965@hiroshima-u.ac.jp (K.K.); m171023@hiroshima-u.ac.jp (X.Y.)
[2] Faculty of Human Culture and Science, Prefectural University of Hiroshima, Hiroshima 734-8558, Japan; takuya62@pu-hiroshima.ac.jp
[3] Department of Fisheries, Cebu Technological University-Carmen Campus, 6005 Cebu, Philippines; glogdelan@gmail.com
[4] Department of Ecology and Evolutionary Biology, University of California, Los Angeles 90095, CA, USA; tomano@ucla.edu
[5] Faculty of Healthcare, Tokyo Health Care University, Tokyo 154-8568, Japan; t-noguchi@thcu.ac.jp
[6] Takehara Station, Setouchi Field Science Center, Graduate School of Biosphere Science, Hiroshima University, Takehara City, Hiroshima 725-0024, Japan; ohtsuka@hiroshima-u.ac.jp
* Correspondence: asakawa@hiroshima-u.ac.jp; Tel.: +81-(0)82-424-7930

Received: 1 April 2019; Accepted: 27 April 2019; Published: 29 April 2019

Abstract: The toxicity of the greater blue-ringed octopus *Hapalochlaena lunulata*, whose bite is fatal to humans, was examined to better understand and prevent deaths from accidental bites. Living specimens were collected from tide pools on Ishigaki Island, Okinawa Prefecture, Japan, in November and December of 2015, 2016, and 2017. The specimens were examined for the anatomical distribution of the toxicity, which was expressed in terms of mouse units (MU), by the standard bioassay method for tetrodotoxin (TTX) in Japan. Paralytic toxicity to mice was detected in all of the soft parts. The posterior salivary glands exhibited the highest toxicity score with a maximum level of 9276 MU/g, which was classified as "strongly toxic" (more than 1000 MU/g tissue) according to the classification of toxicity established by the Ministry of Health, Labor and Welfare of Japan, followed by the hepatopancreas (21.1 to 734.3 MU/g), gonads (not detectable to 167.6 MU/g), arms (5.3 to 130.2 MU/g), and other body areas (17.3 to 107.4 MU/g). Next, the toxin from the salivary glands was partially purified by a Sep-Pak C18 cartridge and an Amicon Ultra Centrifugal Filter with a 3000-Da cut-off, and analyzed by liquid chromatography-mass spectrometry (LC-MS) equipped with a $\varphi 2.0 \times$ 150-mm (5 µm) TSKgel Amide-80 column (Tosoh, Tokyo, Japan) with a mixture of 16 mM ammonium formate buffer (pH 5.5) and acetonitrile (ratio 3:7, *v/v*) as a mobile phase. This study aimed to clarify the toxicity and the composition of TTX and its derivatives in this toxic octopus. The main toxin in this toxic octopus was identified as TTX, along with 4-*epi* TTX, 4, 9-anhydroTTX and 6-*epi* TTX. Further, the toxicity of this species is also significant from a food hygiene point of view.

Keywords: greater blue-ringed octopus; *Hapalochlaena lunulata*; posterior salivary gland; paralytic toxicity; Ishigaki Island; tetrodotoxin; LC-MS

Key Contribution: To the best of our knowledge; this is the first report on the toxicity and toxin composition of *H. lunulata* from Japan.

1. Introduction

It has been known for a long time that several species of octopus secrete from the posterior salivary glands a substance that is toxic to prey organisms [1,2]. In humans, the bites of several species of octopuses can cause pain around the wound area [3–5], and occasionally even fatal lesions. Symptoms of such lesions include numbness of the face as well as acute and progressive skeletal muscle weakness due to the venom from the posterior salivary glands, which is connected to the beak. The posterior salivary glands are the typical storage sites of octopus venoms. Venomous bites by octopuses belonging to the genus *Hapalochlaena* (Cephalopoda: Octopoda: Octopodidae) are among the most dangerous octopus bites, and there have been several reported fatalities and near fatalities resulting from their bites. In Australia, many case reports on bites by toxic octopuses have been recorded [6–12]. Fortunately, bites by this species have not been reported in Japan. The potent neurotoxin contained in the posterior salivary glands of *Hapalochlaena maculosa* (formerly *Octopus maculosus*) that is responsible for fatal bites has been identified as tetrodotoxin (TTX), formerly known as maculotoxin [13–17]. Consequently, TTX have become the major point of focus on cephalopod toxinological research. It is well-known that TTX is an extremely potent low-molecular weight neurotoxin ($C_{11}H_{17}N_3O_8$; MW = 319) that is associated with neurotoxic marine poisonings; it has powerful pharmacological action to block the specific voltage-dependent sodium channels of biological membranes [18]. TTX is one of the most powerful marine biotoxins, and similar to saxitoxin, which is a paralytic shellfish poison, it has a 50% lethal dose (LD_{50}) in mice of 10 µg/kg, as compared to 10,000 µg/kg for sodium cyanide [19]. TTX is a heterocyclic guanide compound whose chemical structure has been characterized. The TTX analogs isolated from puffers can be classified into four groups: (1) analogs chemically equivalent to TTX (4-*epi*TTX and 4,9-anhydroTTX); (2) deoxy analogs (5-deoxyTTX, 11-deoxyTTX, etc.); (3) 11-CH$_2$OH oxidized analogs (11-oxoTTX); and (4) C11-lacking analogs (11-norTTX-6(*S*)-ol and 11-norTTX-6(*R*)-ol) [20,21]. The pharmacological properties of these TTX analogs were revealed to be closely similar to those of TTX.

Human intoxication with this toxin is characterized by symptoms of respiratory failure that result in death in the most severe cases, e.g., TTX poisoning cases due to ingestion of puffer fish. Since there is no cure for TTX poisoning, the mortality rate is very high. The genus *Hapalochlaena* consists of three toxic species: the blue-lined octopus *H. fasciata*, the lesser blue-ringed octopus *H. maculosa*, and the greater blue-ringed octopus *H. lunulata*, all of which are distributed in the tropical to subtropical zones of the Indo-West Pacific, and are especially common in marine waters around Australia. The geographical distribution of *H. lunulata* extends to Indonesia, the Philippines, Papua New Guinea, Vanuatu, the Solomon Islands, and even Japan [22]. Due to its high level of TTX, the greater blue-ringed octopus *H. lunulata* is regarded as one of the most venomous marine animals in the world. However, the toxicity and toxin composition of the greater blue-ringed octopus *H. lunulata* in Japan remain unknown. There is an urgent need to clarify the toxicity and toxin composition of octopuses belonging to the genus *Hapalochlaena* (Cephalopoda: Octopoda: Octopodidae) in Japan, which prompted us to undertake the present study. From the saliva or salivary glands, many physiologically active substances have been separated as salivary gland toxins in cephalopods [2], but this study aimed to clarify the toxicity and the composition of TTX and its derivatives in the greater blue-ringed octopus from Ishigaki Island, Okinawa Prefecture, Japan, as part of a series of studies on the toxification mechanism of TTX-bearing octopuses.

2. Results and Discussion

Table 1 shows the anatomical distribution of the toxicity of the octopus. In the mouse bioassay for lethal potency described below, all seven specimens showed paralytic toxicity, irrespective of the date of collection. The posterior salivary glands exhibited the highest toxicity score with a maximum level of 9276 mouse units (MU)/g (total toxicity per specimen, 234.8 MU), which was classified as "strongly toxic" (more than 1000 MU/g tissue) according to the classification of toxicity established by the Ministry of Health, Labor and Welfare of Japan, followed by the hepatopancreas (21.1 to

734.3 MU/g), gonads (not detectable to 167.6 MU/g), other body parts (17.3 to 107.4 MU/g), and arms (5.3 to 130.2 MU/g). Due to the high levels of TTX, the greater blue-ringed octopus *H. lunulata* is regarded as one of the most venomous marine animals. The toxicity of the whole body was assessed in two specimens; they exhibited a toxicity of 74.3 and 100.4 MU/g, and the range of total toxicity per specimen was 61.7 to 234.8 MU/g. These findings suggest that among all of the body parts of *H. lunulata* in Japan, the posterior salivary glands have the highest toxicity, although the toxicity was not exclusively localized in this organ. In contrast, when the toxicity of the posterior salivary glands and other different body parts were examined in another study using specimens of *H. maculosa* collected in the Philippines [23], the posterior salivary glands exhibited a toxicity of 274 MU/g (total toxicity per specimen, 41 MU) while the other body parts exhibited a toxicity of 11 MU/g (total toxicity per specimen, 133 MU); in other words, the toxin was distributed mainly in the other soft body parts, and not in the posterior salivary glands. In the present study, there were wide inter-specimen variations in the toxicity of the greater blue-ringed octopus *H. lunulata*, and correct identification of this species is difficult. The misidentification of *H. lunulata* is a potential risk to human safety as it may lead to fatal bites and severe morbidity, especially if it is misidentified as an edible octopus. In fact, food poisoning incidents due to the ingestion of TTX-bearing octopus *H. fasciata* have been reported in Taiwan [24].

Table 1. Anatomical distribution of the toxicity of the greater blue-ringed octopus *Hapalochlaena lunulata* from Ishigaki Island, Okinawa Prefecture, Japan.

Year	2015					2016		
Date of Collection	Nov.28			Dec.27		Nov.17		
Weight of whole body (g)	9.63		5.24	0.83		5.30		1.66
Organ	Weight (g)	Toxicity (MU/g)	Weight (g)	Toxicity (MU/g)	Toxicity (MU/g)	Weight (g)	Toxicity (MU/g)	Toxicity (MU/g)
Posterior salivary glands	0.37	288.0	0.02	9276		0.04	704.9	
Gonad	0.05	52.3	0.03	ND	74.3 *	0.13	-	100.4 *
Hepatopancreas	0.38	58.8	0.17	145.4		0.42	21.1	
Arm	3.28	5.3	1.99	9.0		2.85	7.5	
Others	0.60	26.3	0.36	17.3		1.86	68.2	
Total toxicity/sp.	163.9		234.8		61.7	185		167.0

Year	2017							
Date of Collection	Dec.4							
Weight of whole body (g)	1.08		3.14		3.65		0.71	
Organ	Weight (g)	Toxicity (MU/g)	Weight (g)	Toxicity (MU/g)	Weight (g)	Toxicity (MU/g)	Weight (g)	Toxicity (MU/g)
Posterior salivary glands	0.02	1729	0.02	1059	0.04	491.1	0.003	5002.9
Gonad	0.02	ND	0.04	167.6	0.06	ND	0.02	ND
Hepatopancreas	0.11	265.4	0.35	195.3	0.15	88.7	0.04	734.3
Arm	0.49	10.7	1.89	21.7	1.07	22.2	0.28	130.2
Others	0.44	21.6	0.84	107.4	2.03	20.7	0.16	80.2
Total toxicity/sp.	78.5		227.0		98.7		93.7	

-: not tested; ND: not detected; *: whole body.

In this context, in 2016 and 2018, blue-lined octopus *H. fasciata* specimens were collected in Japan, one in Lake Hamana (brackish water lake), Shizuoka Prefecture, and one off the coast of Boso Peninsula, Chiba Prefecture [25]. Although the toxicity of the posterior salivary glands was below the level of detection in one octopus and 42.5 MU/g in the other octopus, the fact that *Hapalochlaena* spp. were found in the coastal waters of Japan proper within the temperate zone may indicate that TTX-bearing octopuses could become a serious health issue in Japan in the near future. Further, the toxicity of these species is also significant from a food hygiene point of view.

In humans, the minimum lethal dose of TTX is estimated to be approximately 10,000 MU (1 MU = 0.178 µg), which is equivalent to 2 mg of TTX crystal [26]. TTX is heat-resistant and does not decompose during general cooking processes, such as heating and boiling, and there are presently no known antidotes or antitoxins to TTX. Therefore, treatments for TTX poisoning are considered to be only supportive. TTX poisoning is characterized by a few symptoms in the victims. The symptoms depend on the amount of toxin ingested as well as the age and health of the victim. In Australia, the fatal envenomation of adult green sea turtles by accidental consumption of seagrass and blue-lined octopuses together has also been reported [27].

Figure 1. shows the results of liquid chromatography-mass spectrometry (LC-MS) analysis. The ion-monitored mass chromatograms show protonated molecular ion peaks (M + H)$^+$ at *m/z* = 320 and 302, which coincided well with data for TTX standards [20,28,29]. TTX and two TTX analogs (4-*epi*TTX and 6-*epi*TTX) all had the same molecular weight (319 Da). The protonated molecular ion peak (M + H)$^+$ of 6-*epi* TTX estimated from the literature [29] was also detected. The main component was identified as TTX along with 4-*epi* TTX, 4,9-anhydroTTX, and 6-*epi*TTX. It could be unambiguously concluded from the symptoms in the mice and the results of LC-MS analysis that the toxins contained in *H. lunulata* collected on Ishigaki Island, Okinawa Prefecture, Japan, comprise a mixture of TTX and TTX derivatives. High-resolution LC-MS/MS analysis is also useful to ensure identification of targets [30]. Multiple reaction monitoring (MRM) mass spectral analysis was not examined this time. As for toxins contained in *Hapalochlaena* sp., existence of peptide and protein neurotoxins are reported [31]. These points were not examined this time.

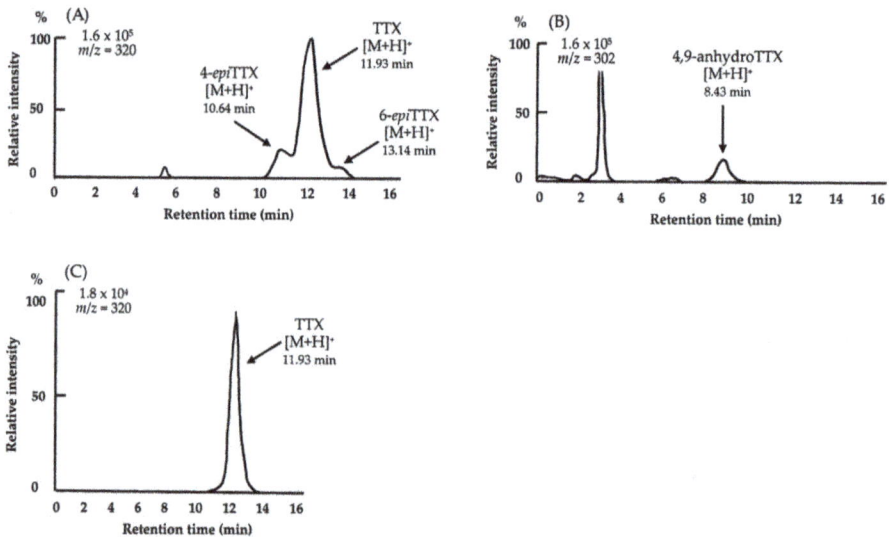

Figure 1. Selected ion-monitored liquid chromatography-mass spectrometry (LC-MS) chromatograms of the toxin from the posterior salivary glands of the greater blue-ringed octopus *Hapalochlaena lunulata*. (**A**,**B**) Toxins from the posterior salivary glands. (**A**) *m/z* = 320; (**B**) *m/z* = 302; (**C**) Reference standard samples of TTX; *m/z* = 320.

To the best of our knowledge, this is the first report on the toxicity and toxin composition of *H. lunulata* from a subtropical area of Japan. In another study using post-column fluorescence high-performance liquid chromatography, TTX was present in all of the body parts of *H. maculosa* from South Australia, including high concentrations of TTX in the arms, abdomen, and cephalothorax [32]. In contrast, TTX was found only in the posterior salivary glands, mantle tissue and ink of *H. lunulata* from Bali, Indonesia [33]. Octopuses generally have an ink sac. When they encounter an enemy or are attacked by them, they secrete a lot of ink from the sac to help conceal themselves to escape. While TTX is generally accepted to be a powerful chemical defense against predators [19,34,35], offensive functions have also been suggested for the release of TTX. Octopuses are carnivores, and their salivary glands, especially the posterior salivary glands, produce venom to assist them in capturing various crustacean preys, such as shrimps and crabs. In blue-ringed octopuses, TTX may serve as a hunting tool for paralyzing prey as well as a biological defense tool against predation for eggs and young blue-ringed octopuses of planktonic stage [36]. Thus, TTX-bearing octopuses appear to use TTX for capturing prey as well as for protecting themselves from enemies. On the other hand, TTX levels of adult females, paralarvae, and eggs were investigated to ascertain the relationship between maternal and offspring TTX levels, and to examine TTX-ontology through hatching. It is suggested that embryos or their bacterial symbionts begin independent production of TTX before hatching [36]. In this connection, it has demonstrated that TTX levels in the embryos of puffer fish increase until hatching; emphasizing its endogenous origin [37].

TTX is found in a remarkably wide range of marine and terrestrial animals across disparate taxa; it is found not only in pufferfish, but also in a variety of vertebrates and invertebrates [38–40]. It is generally accepted that TTX is accumulated in TTX-bearing animals through the food chain, starting from bacteria as the ultimate producers of the toxin, although the exact biosynthetic and metabolic pathways of TTX remain unknown. However, it is also possible that TTX is not obtained via the food chain and is instead produced by symbiotic or parasitic bacteria that directly accumulate inside of the octopuses. It is not clear at present whether the toxins in our toxic octopus specimens were endogenous or exogenous in origin. Since octopuses are generally carnivorous feeders, it is more plausible that octopus specimens accumulate the toxin by feeding on toxic marine organisms in the sampling areas. It is not uncommon for toxins to be transported and accumulated in food chains, particularly in marine biota, and feeding experiments may be useful for clarifying the origins of the toxins contained in toxic octopuses. Evidence of TTX-producing bacteria isolated from some TTX-bearing organisms, such as puffer fish and xanthid crabs, exists [41,42]. Although the origin of TTX in the venomous octopuses in this study remains unclear, TTX appears to be produced by bacteria in the posterior salivary glands of *H. maculosa* [23]. In our study, wide inter-specimen variations in toxicity were observed even within the same species (Table 1), suggesting that the level of TTX in toxic octopuses is related to some environmental factors, or that it comes from food. Nonetheless, the mechanism may involve factors other than the food chain. Further research to elucidate the associated mechanisms of toxification is now in progress. In addition, investigations on specimen-, location-, and size-dependent variations in the toxicity of *H. lunulata* are also needed, and results in comparison to those of the present study will be published elsewhere at a later date.

3. Materials and Methods

3.1. Materials

Figure 2 shows the sampling location off the coast of Ohsaki beach in Ishigaki Island, Okinawa Prefecture, Japan. A total of nine specimens of this octopus, which were found swimming in tide pools, were collected with a small type of landing net. In this sampling area, tide pools suitable for the sampling of these species appeared around midnight in the winter seasons throughout 2015 to 2017. Figure 3 shows a representative specimen of the greater blue-ringed octopus *H. lunulata* that was assessed in this study. This octopus was identified by its morphological characteristics with reference to an illustrated catalog of cephalopod species [22]. The total body length was around 10 cm from the mantle apex to the arm tips. The appearance of this live octopus was characterized by numerous small brilliant blue rings, which it flashed as an aposematic warning signal, scattered on its arms and body. Dr. T. Okutani (Prof. Emeritus, Tokyo University of Marine Science and Technology), a taxonomist of cephalopods, kindly confirmed our species identification. The animals were placed in individual 50 mL polypropylene conical tubes filled with fresh seawater, and transported alive by air from Ishigaki Island to the Laboratory of Marine Bioresource Chemistry, Hiroshima University. After the seawater was discarded, the animals were frozen and stored at –20 °C for no longer than 1 month before analysis.

Following identification according to the anatomical location of a pair of posterior salivary glands with reference to previous reports [2,31,43,44], the posterior salivary glands were carefully excised.

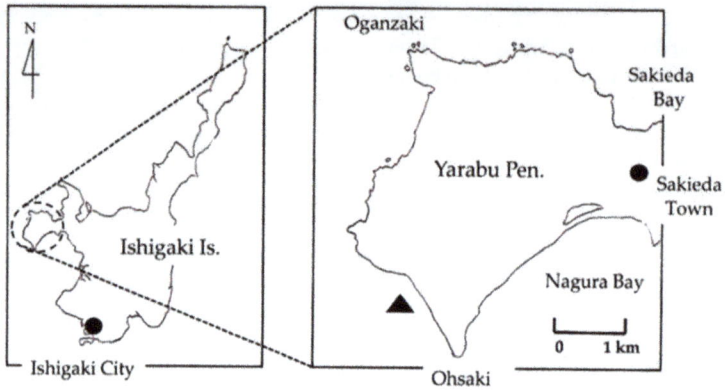

Figure 2. Map showing the location for the collection of greater blue-ringed octopus *Hapalochlaena lunulata* on Ishigaki Island (N24°20′, E124°09′), Okinawa Prefecture, Japan. Ishigaki Island is shown in the map on the left. The map on the right is an enlarged image of Yarabu Peninsula showing the sampling location (▲). ●: Ishigaki City in the map on the left and Sakieda Town in the map on the right.

(A)   (B)

Figure 3. Greater blue-ringed octopus *Hapalochlaena lunulata* from Ishigaki Island, Okinawa Prefecture, Japan. (**A**) Living whole body. White scale bar = 1.0 cm. (**B**) Location of the posterior salivary glands (yellow arrows).

3.2. Mouse Bioassay for Lethal Potency

To examine the anatomical distribution of toxicity, the specimens were dissected into five parts: the posterior salivary glands, gonads, hepatopancreas, arms, and other body parts. Because the principal toxin was suspected to be TTX, the standard bioassay method for TTX [45] was used with slight modifications. Briefly, the tissues were finely cut with scissors, ground with a mortar and pestle, and then transferred into a glass test tube containing 3 mL of 0.1% acetic acid. The mixture was heated in a boiling water bath for 5 min, cooled, then centrifuged at 11,000× g for 10 min at 4 °C. One milliliter of the supernatant or its dilution was intraperitoneally injected into male mice of the ddY strain (18 to

20 g in body weight). Lethality was expressed in terms of MU, where 1 MU was defined as the amount of toxin that kills a mouse in 30 min.

3.3. Preparation and Identification of Toxins

The remaining extracted liquid for the mouse bioassay was evaporated until dry, and the sample was then dissolved in a small amount of water. Each sample was centrifuged, and the supernatant was applied to a Sep-Pak C18 cartridge (Waters, Milford, MA, USA) equilibrated with water after washing with MeOH. The unbound toxic fraction was concentrated, freeze-dried, dissolved in a small amount of water, and then ultra-filtered in an Amicon Ultra Centrifugal Filter with a 3000-Da cut-off (Merck Millipore, Cork, Ireland) by centrifugation at 5300× *g* for 15 min at 4 °C. The clear filtrate was used as the sample solution and was subjected to analysis by an electrospray ionization-liquid chromatography-mass spectrometry (ESI-LC/MS) system according to the method of Nakagawa et al. [29]. For LC-MS, an LC system (Agilent 1100 series; Agilent Technologies, Palo Alto, CA, USA) with a φ2.0 × 150 mm (5 μm) TSKgel Amide-80 column (Tosoh, Tokyo, Japan) was used with a mixture of 16 mM ammonium formate buffer (pH 5.5) and acetonitrile (ratio 3:7, *v/v*) as a mobile phase at a flow rate of 0.2 mL/min at 25 °C. The column was connected to the electrospray interface of an API2000 quadrupole mass spectrometer (Applied Biosystems, Warrington, UK). Six ions at *m/z* 272 (5,6,11-trideoxyTTX), 288 (5,11-dideoxyTTX, 6,11-dideoxyTTX), 290 (11-*nor*TTX-6(*S*)-ol), 302 (4,9-anhydroTTX), 304 (5-deoxyTTX, 11-deoxyTTX) and 320 (4-*epi*TTX, 6-*epi*TTX, TTX), corresponding to the [M + H]$^+$ of TTX and its analogs, were detected in selected ion-monitoring (SIM) mode. The elution time of the TTX analogs in the partially purified *H. lunulata* toxins was calculated based on the TTX standards with reference to previously reported elution times [20,28,29]. Reference standard samples of TTX were essentially prepared by chromatography on activated charcoal, Bio-Gel P-2, and Bio-Rex 70 (H$^+$ form) from ribbon worm *Cephalothrix simula*, as reported previously [38].

Author Contributions: This research work was conceived and designed by M.A.; M.A., T.M., K.U., K.K., X.Y. and S.T. organized and performed the experiments. M.A., T.M., G.G.-D., S.T., S.O. and T.N. interpreted the data. All authors have read and approved the final manuscript.

Funding: This work was supported by Japan Society for the Promotion of Science, Grant-in-Aid for Scientific Research, Grant Numbers 15K07577 and 16K07825, awarded to Manabu Asakawa and Susumu Ohtsuka, respectively.

Acknowledgments: We are grateful to Kotaro Tsuchiya of the Tokyo University of Marine Science and Technology, and Tsunemi Kubodera of the National Museum of Nature and Science for useful information on blue-ringed octopuses (genus *Hapalochlaena*).

References

1. Ghiretti, F. Toxicity of octopus saliva against Crustacea. *Ann. N.Y. Acad. Sci.* **1960**, 726–741. [CrossRef]
2. Hashimoto, Y. Salivary gland toxins in cephalopods. In *Marine Toxins and Other Bioactive Metabolites*; Hashimoto, Y., Ed.; Japan Scientific Societies Press: Tokyo, Japan, 1979; pp. 189–194.
3. Ito, H.; Yoshiba, S.; Honda, M.; Niimura, M. Two cases of octopus bite and examination on *Hapalochlaena* bite. *Jpn. J. Clinic. Dermatol.* **2000**, *54*, 9–13.
4. Ishina, K.; Nakagawa, K.; Ishii, M. *Octopus vulgaris* bite. *Jpn. J. Clinic. Dermatol.* **2008**, *54*, 9–13.
5. Haddad, V., Jr; de Magalhães, C.A. Infiltrated plaques resulting from an injury caused by the common octopus (*Octopus vulgaris*): A case report. *J. Venom. Toxins incl. Trop. Dis.* **2014**, *20*. [CrossRef] [PubMed]
6. Flecker, H.; Cotton, B.C. Fatal bite from octopus. *Med. J. Aust.* **1955**, *42*, 329–331.
7. Hopkins, D.G. Venomous effects and treatment of octopus bite. *Med. J. Aust.* **1964**, *1*, 81–82.
8. Lane, W.R.; Sutherland, S. The ringed octopus bite: A unique medical emergency. *Med. J. Aust.* **1967**, *2*, 475–476.
9. Hodgson, W.C. Pharmacological action of Australian animal venoms. *Clin. Exp. Pharmacol. Physiol.* **1997**, *24*, 10–17. [CrossRef]

10. Cavazzoni, E.; Lister, B.; Sarget, O.; Schiber, A. Blue-ringed octopus (*Hapalochlaena* sp.) envenomation of a 4-year-old boy: A case report. *Clinic. Toxicol.* **2008**, *46*, 760–761. [CrossRef]
11. Jacups, S.P.; Currie, B.J. Blue-ringed octopuses: A brief review of their toxicology. *North. Territ. Nat.* **2008**, *20*, 50–57.
12. McMichael, D.F. The identity of the venomous octopus responsible for a fatal bite at Darwin, Northern Territory. *J. Malacol. Soc. Aust.* **1964**, *1*, 23–24. [CrossRef]
13. Jarvis, M.W.; Crone, H.D.; Freeman, S.E.; Turner, B.J.; Turner, R.J. Chromatographic properties of maculotoxin, a toxin secreted by octopus (*Hapalochlaena*) maculosa. *Toxicon* **1975**, *13*, 177–181. [CrossRef]
14. Crone, H.D.; Leake, B.; Jarvis, M.W.; Freeman, S.E. On the nature of "Maculotoxin", a toxin from blue-ringed octopus (*Hapalochlaena maculosa*). *Toxicon* **1976**, *14*, 423–426. [CrossRef]
15. Sheumack, D.D.; Howden, M.E.; Spence, I.; Quinn, R.J. Maculotoxin: A neurotoxin from the venom glands of the octopus *Hapalochlaena maculosa* identified as tetrodotoxin. *Science* **1978**, *199*, 188–189. [CrossRef]
16. Sheumack, D.D.; Howden, M.E.H.; Spence, L. Occurrence of a tetrodotoxin-like compound in the eggs of the venomous blue-ringed octopus (*Hapalochlaena maculosa*). *Toxicon* **1984**, *22*, 811–812. [CrossRef]
17. Bonnet, M.S. The toxicology of *Octopus maculosa*: The blue-ringed octopus. *Br. Hom. J.* **1999**, *88*, 166–171. [CrossRef]
18. Narahashi, T. Pharmacology of tetrodotoxin. *J. Toxicol.: Toxin Rev.* **2001**, *20*, 67–84. [CrossRef]
19. Mosher, H.S.; Fuhrman, F.A.; Buchwald, H.D.; Fischer, H.G. Tarichatoxin-tetrodotoxin: A potent neurotoxin. *Science* **1964**, *144*, 1100–1110. [CrossRef]
20. Yotsu-Yamashita, M.; Abe, Y.; Kudo, Y.; Ritson-Williams, R.; Paul, V.J.; Konoki, K.; Cho, Y.; Adachi, M.; Imazu, T.; Nishikawa, T.; et al. First identification of 5,11-dideoxytetrodotoxin in marine animals, and characterization of major fragment ions of tetrodotoxin and its analogs. *Mar. Drugs.* **2013**, *11*, 2799–2813. [CrossRef]
21. Ueyama, N.; Sugimoto, K.; Kudo, Y.; Onodera, K.; Cho, Y.; Konoki, K.; Nishikawa, T.; Yotsu-Yamashita, M. Spiro bicyclic guanidino compounds from pufferfish: Possible biosynthetic intermediates of tetrodotoxin in marine environments. *Chem.—Eur.J.* **2018**, *24*, 7250–7258. [CrossRef]
22. Norman, M.D.; Finn, J.K.; Hochberg, F.G. Family Octopodidae. In *Cephalopods of the World, an Annotated and Illustrated Catalogue of Cephalopods Species Known to Date, FAO Species Catalogue for Fishery Purposes*; Jereb, P., Roper, C.F.E., Norman, M.D., Finn, J.K., Eds.; Food and Agriculture Organization of the United Nations: Rome, Italy, 2016; Volume 4, pp. 136–140.
23. Hwang, D.F.; Arakawa, O.; Saito, T.; Noguchi, T.; Simidu, U.; Tsukamoto, K.; Shida, Y.; Hashimoto, K. Tetrodotoxin-producing bacteria from the blue-ringed octopus *Octopus maculosa*. *Mar. Biol.* **1989**, *100*, 327–332. [CrossRef]
24. Wu, Y.-J.; Lin, C.-L.; Chen, C.-H.; Hsieh, C.-H.; Jen, H.-C.; Jian, S.-J.; Hwang, D.-F. Toxin and species identification of toxic octopus implicated into food poisoning in Taiwan. *Toxicon* **2014**, *91*, 96–102. [CrossRef]
25. Asakawa, M.; Umezaki, K.; Fujii, M.; et al. Toxicity and toxin composition of blue-lined octopus H. fasciata in Japan. In Proceedings of the 114th Meeting of Japanese Society for Food Hygiene and Safety, Hiroshima, Japan, 15–16 November 2018.
26. Hwang, D.F.; Noguchi, T. Tetrodotoxin poisoning. *Adv. Food. Nutr. Res.* **2007**, *52*, 141–236. [PubMed]
27. Townsend, K.A.; Altvater, J.; Thomas, M.C.; Schuyler, Q.A.; Nette, G.W. Death in the octopus' garden: Fatal blue-lined octopus envenomations of adult green sea turtles. *Mar. Biol.* **2012**, *159*, 689–695. [CrossRef]
28. Arakawa, O.; Noguchi, T.; Shida, Y.; Onoue, Y. Occurrence of 11-*oxo*tetrodotoxin and 11-*nor*tetrodotoxin-6(R)-ol in a xanthid crab *Atergatis floridus* collected at Kojima, Ishigaki Island. *Fish. Sci.* **1994**, *60*, 769–771. [CrossRef]
29. Nakagawa, T.; Jang, J.; Yotsu-Yamashita, M. Hydrophilic interaction liquid chromatography-electrospray ionization mass spectrometry of tetrodotoxin and its analogs. *Anal. Biochem.* **2006**, *352*, 142–144. [CrossRef] [PubMed]
30. Bane, V.; Hutchinson, S.; Sheehan, A.; Brosnan, B.; Barnes, P.; Lehane, M.; Furey, A. LC-MS/MS method for the determination of tetrodotoxin (TTX) on a triple quadruple mass spectrometer. *Food Addit. Contam.: Part A* **2016**, *33*, 1728–1740. [CrossRef]
31. Fry, B.G.; Roelants, K.; Norman, J.A. Tentacles of venom: Toxic protein convergence in the kingdom Animalia. *J. Mol. Evol.* **2009**, *68*, 311–321. [CrossRef]
32. Yotsu-Yamashita, M.; Mebs, D.; Flachsenberger, W. Distribution of tetrodotoxin in the body of the blue-ringed octopus (*Hapalochlaena maculosa*). *Toxicon* **2007**, *49*, 410–412. [CrossRef]

33. Williams, B.L.; Caldwell, R.L. Intra-organismal distribution of tetrodotoxin in two species of blue-ringed octopuses (*Hapalochlaena fasciata* and *H. lunulata*). *Toxicon* **2009**, *54*, 345–353. [CrossRef]

34. Miyazawa, K.; Jeon, J.K.; Noguchi, T.; Ito, K.; Hashimoto, K. Distribution of tetrodotoxin in the tissues of the flatworm *Planocera multitentaculata* (Platyhelminthes). *Toxicon* **1987**, *25*, 975–980. [CrossRef]

35. Kodama, M.; Ogata, T.; Sato, S. External secretion of tetrodotoxin from puffer fishes stimulated by electric shock. *Mar. Biol.* **1985**, *87*, 199–202. [CrossRef]

36. Williams, B.L.; Hanifin, C.T.; Brodie, E.D.; Caldwell, R.L. Ontogeny of tetrodotoxin levels in blue-ringed octopuses: Maternal investment and apparent independent production in offspring of *Hapalochlaena lunulata*. *J. Chem. Ecol.* **2011**, *37*, 10–17. [CrossRef] [PubMed]

37. Matsumura, K. Production of tetrodotoxin in puffer fish embryos. *Environ. Toxicol. Pharmacol.* **1998**, *6*, 217–219. [CrossRef]

38. Asakawa, M.; Toyoshima, T.; Ito, K.; Bessho, K.; Yamaguchi, C.; Tsunetsugu, S.; Shida, Y.; Kajihara, H.; Mawatari, F.S.; Noguchi, T.; et al. Paralytic toxicity in the ribbon worm *Cephalothrix* species (Nemertean) in Hiroshima Bay, Hiroshima Prefecture, Japan and the isolation of tetrodotoxin as a main component of its toxins. *Toxicon* **2003**, *41*, 747–753. [CrossRef]

39. Noguchi, T.; Arakawa, O. Tetrodotoxin-distribution and accumulation in aquatic organisms, and cases of human intoxication. *Mar. Drugs.* **2008**, *6*, 220–242. [CrossRef]

40. Asakawa, M.; Ito, K.; Kajihara, H. Highly toxic ribbon worm *Cephalothrix simula* containing tetrodotoxin in Hiroshima Bay, Hiroshima Prefecture, Japan. *Toxins* **2013**, *5*, 376–395. [CrossRef] [PubMed]

41. Noguchi, T.; Jeon, J.-K.; Arakawa, O.; Sugita, H.; Deguchi, Y.; Shida, Y.; Hashimoto, K. Occurrence of tetrodotoxin and anhydrotetrodotoxin in *Vibrio* sp. isolated from the intestines of a xanthid crab, *Atergatis floridus*. *J. Biochem.* **1986**, *99*, 311–314. [CrossRef]

42. Noguchi, T.; Hwang, D.F.; Arakawa, O.; Sugita, H.; Deguchi, Y.; Shida, Y.; Hashimoto, K. *Vibrio alginolyticus*, a tetrodotoxin-producing bacterium in the intestines of the fish Fugu *vermicularis vermicularis*. *Mar. Biol.* **1987**, *94*, 625–630. [CrossRef]

43. Gibbs, P.J.; Greenaway, P. Histological structure of the posterior salivary glands in the blue ringed octopus *Hapalochlaena maculosa* Hoyle. *Toxicon* **1978**, *16*, 59–70. [CrossRef]

44. Allcock, A.L.; Hochberg, F.G.; Rodhouse, P.G.K.; Thorpe, J.P. *Adelieledone*, a new genus of octopodid from the Southern Ocean. *Antarct. Sci.* **2003**, *15*, 415–424. [CrossRef]

45. Kawabata, T. *Food Hygiene Examination Manual*; Food Hygiene Association: Tokyo, Japan, 1978; pp. 232–240.

toxins

MDPI

Article

Paralytic Shellfish Toxins in Surf Clams *Mesodesma donacium* during a Large Bloom of *Alexandrium catenella* Dinoflagellates Associated to an Intense Shellfish Mass Mortality

Gonzalo Álvarez [1,2,*], **Patricio A. Díaz** [3], **Marcos Godoy** [4,5,6,*], **Michael Araya** [2], **Iranzu Ganuza** [1], **Roberto Pino** [1,6], **Francisco Álvarez** [1], **José Rengel** [1], **Cristina Hernández** [7], **Eduardo Uribe** [1] and **Juan Blanco** [8]

1   Facultad de Ciencias del Mar, Departamento de Acuicultura, Universidad Católica del Norte, Larrondo 1281, Coquimbo 1781421, Chile; iranzuganu@gmail.com (I.G.); roberto.pino@alumnos.ucn.cl (R.P.); falvarezsego@gmail.com (F.Á.); jose.rengel@ucn.cl (J.R.); euribe@ucn.cl (E.U.)
2   Centro de Investigación y Desarrollo Tecnológico en Algas (CIDTA), Facultad de Ciencias del Mar, Universidad Católica del Norte, Larrondo 1281, Coquimbo, Chile; mmaraya@ucn.cl
3   Centro i~mar & CeBiB, Universidad de Los Lagos, Casilla 557, Puerto Montt 5480000, Chile; patricio.diaz@ulagos.cl
4   Laboratorio de Biotecnología Aplicada, Facultad de Ciencias Veterinarias, Universidad San Sebastián, Lago Panguipulli 1390, Puerto Montt 5501842, Chile
5   Centro de Investigaciones Biológicas Aplicadas (CIBA), Diego de Almagro 1013, Puerto Montt 5507964, Chile
6   Doctorado en Acuicultura, Programa Cooperativo Universidad de Chile, Universidad Católica del Norte, Pontificia Universidad Católica de Valparaíso, Coquimbo 17811421, Chile
7   Laboratorio Salud Pública, Seremi de Salud Región de Los Lagos, Crucero 1915, Puerto Montt 5505081, Chile; cristina.hernandez@redsalud.gov.cl
8   Centro de Investigacións Mariñas (Xunta de Galicia), Apto. 13, 36620 Vilanova de Arousa, Pontevedra, Spain; juan.carlos.blanco.perez@xunta.gal
*   Correspondence: gmalvarez@ucn.cl (G.Á.); marcos.godoy@ciba.cl (M.G.); Tel.: +56-51-2209766 (G.Á.)

Received: 8 March 2019; Accepted: 27 March 2019; Published: 29 March 2019

Abstract: In late February 2016, a harmful algal bloom (HAB) of *Alexandrium catenella* was detected in southern Chiloé, leading to the banning of shellfish harvesting in an extended geographical area (~500 km). On April 24, 2016, this bloom produced a massive beaching (an accumulation on the beach surface of dead or impaired organisms which were drifted ashore) of surf clams *Mesodesma donacium* in Cucao Bay, Chiloé. To determine the effect of paralytic shellfish poisoning (PSP) toxins in *M. donacium*, samples were taken from Cucao during the third massive beaching detected on May 3, 2016. Whole tissue toxicity evidence a high interindividual variability with values which ranged from 1008 to 8763 μg STX eq 100 g^{-1} and with a toxin profile dominated by GTX3, GTX1, GTX2, GTX4, and neoSTX. Individuals were dissected into digestive gland (DG), foot (FT), adductor muscle (MU), and other body fractions (OBF), and histopathological and toxin analyses were carried out on the obtained fractions. Some pathological conditions were observed in gill and digestive gland of 40–50% of the individuals that correspond to hemocyte aggregation and haemocytic infiltration, respectively. The most toxic tissue was DG (2221 μg STX eq 100 g^{-1}), followed by OBF (710 μg STX eq 100 g^{-1}), FT (297 μg STX eq 100 g^{-1}), and MU (314 μg STX eq 100 g^{-1}). The observed surf clam mortality seems to have been mainly due to the desiccation caused by the incapability of the clams to burrow. Considering the available information of the monitoring program and taking into account that this episode was the first detected along the open coast of the Pacific Ocean in southern Chiloé, it is very likely that the *M. donacium* population from Cucao Bay has not had a recurrent exposition to *A. catenella* and, consequently, that it has not been subjected to high selective pressure for PSP resistance. However, more research is needed to determine the effects of PSP toxins on behavioral and physiological responses, nerve sensitivity, and genetic/molecular basis for the resistance or sensitivity of *M. donacium*.

Keywords: *Alexandrium catenella*; PSP outbreak; *Mesodesma donacium*; mass mortality; southern Chile

Key Contribution: Individuals of surf clams *M. donacium* were found with their foot and siphon paralyzed and with high toxicity levels of PST in all tissues during an intense bloom of *Alexandrium catenella*. Hemocyte aggregation and haemocytic infiltration were the main toxicological responses found in *M. donacium* tissues. Surf clam mortality seems to have been mainly due to the desiccation caused by the incapability of the clams to burrow.

---

1. Introduction

Paralytic shellfish poisoning (PSP) is a neurotoxic syndrome caused by the ingestion of shellfish contaminated by saxitoxin and/or its analogues, causing a range of symptoms from slight tingling sensation or numbness around the lips to fatal respiratory paralysis (reviewed by Reference [1]). The toxins involved in this syndrome (paralytic shellfish toxins, PST) are known to be biosynthesized by various species of marine dinoflagellates of the genera *Alexandrium*, *Gymnodinium* and *Pyrodinium* [2–4]. PST include more than 57 structurally related compounds from the marine environment [5], which can be grouped into six classes: (a) N-sulfocarbamoyl toxins (GTX5, GTX6, C1–C4), (b) decarbamoyl toxins (dcGTX1–4, dcNeo, dcSTX), (c) carbamoyl toxins (GTX1–4, neoSTX and STX) [6,7], (d) deoxydecarbamoyl (doSTX, doGTX2–3) [6], (e) hydroxybenzoyl toxins (GC1 to GC6) [8–10], and other saxitoxin analogues that include 11β-hydroxy-N-sulfocarbamoylsaxitoxin (M1), 11β-hydroxysaxitoxin (M2), 11,11-dihydroxy-N-sulfocarbamoylsaxitoxin (M3), 11,11-dihydroxysaxitoxin (M4), and the unidentified compound (M5) [11].

Some HAB species could affect shellfish, with the responses depending on a series of species-specific or individual characteristics of both phytoplankton (including toxin production) and shellfish [12]. Several behavioral, physiological, and cellular responses of shellfish to toxic *Alexandrium* species have been described, including changes in valve closure, filtration rate, feeding rate, byssus production, oxygen consumption, cardiac activity, neurophysiological effects, and pathological alterations [12–25]. High concentrations of organisms belonging to this genus of dinoflagellates and the persistence of its blooms can produce mass mortalities of shellfish, as those reported from different geographical areas and affecting diverse bivalve species, such as mussel *Mytilus meridionalis*, oyster *Crassostrea virginica*, scallop *Chlamys opercularis*, cockle *Cerastoderma edule*, and clams *Donax serra*, *D. variabilis*, and *Spisula solidissima* [12,26–28].

PSP was detected in Chiloé for the first time in 1972, in Bell Bay, Magallanes Region (53°55′ S; 71°45′ W). During that event, three fishermen died due the consumption of the ribbed mussel *Aulacomya atra*, whose toxicity was associated with the presence of the dinoflagellate *Gonyaulax catenella* (currently *Alexandrium catenella*) [29]. Since then, *A. catenella* blooms and PSP toxicity outbreaks have been reported annually covering an extensive area between Magallanes Region to Los Lagos Region, specifically in Quellón, Chiloé Island (43°07′ S; 73°36′ W) [30–32], with interannual differences in the affected geographical area and toxicity intensity [33–38].

In 2016, a late summer bloom of *A. catenella* was detected in Southern Chile. This outbreak was the worst event of all those recorded in Chile in terms of geographical extension and affected species, spreading, for the first time, from southernmost of the Chiloé Archipelago (43°50′ S) to Mehuín, Los Ríos Region (39°25′ S) [39], therefore affecting the Pacific Ocean coastal zone. During the toxic episode, the largest invertebrate mass die-off ever recorded in Chile took place. It included mortalities of different organisms, such as the mollusks *Mesodesma donacium* and *Gari solida* and the crustaceans *Austromegabalanus psittacus* and *Romaleon polyodon* (Figure 1). Furthermore, this toxic episode caused the mortality of vertebrates, including sea lions (*Otaria flavescens*), seagulls (*Larus dominicanus*), and dogs [39]. Human intoxications were also reported (12 people were affected), causing dramatic socioeconomic impacts for more than a thousand fishermen in the affected area [40] because of the banning of shellfish harvesting from the involved natural beds.

Figure 1. Massive beaching of different invertebrate species recorded in Cucao Bay on May 3, 2016, during an *Alexandrium catenella* toxic bloom.

Mesodesma donacium is an endemic species of the Pacific coast of South America, where it is distributed from Sechura Bay, Perú (5° S) to Chiloé Island, Chiloé (43° S) [41,42], inhabiting oceanic sandy beaches, often located near river mouths, that are characterized by strong waves and highly active sediment dynamics [43]. The populations are confined to the subtidal and intertidal zones, where they are buried in the substrate between 10 to 25 cm depth. Natural beds of this species are distributed along the coast in a patchy way [42,44]. In Chiloé, this is one of the most important commercial species for benthic fisheries, especially in the two most important natural beds located in Coquimbo Bay in northern Chiloé and Cucao Bay, where it is the principal economic resource for local fishermen belonging to the Huilliche ethnic group [45,46].

The aim of the current work was to describe the accumulation of PSP toxins, their distribution and profile in different tissues of *M. donacium*, and their possible link with the massive beaching of this species which took place during a large bloom of *Alexandrium catenella*.

2. Results

2.1. Toxicity of Surf Clams and Massive Beaching in Cucao Bay

The analysis of surf clam samples collected from Cucao Bay revealed the occurrence of a PSP toxic episode with toxicities that increased rapidly during the first three sampling weeks (Figure 2). The beginning of the episode (Figure 3) was detected on March 24, 2016, in Chanquín with a toxicity of 33 μg STX eq 100 g^{-1}. Two weeks later on April 6, an abrupt increase of toxicity was detected in all localities reaching values of 394, 351, 428 μg STX eq 100 g^{-1} for Chanquín, Palihue and Deñal, respectively, that exceeded the regulatory limit (80 μg STX eq 100 g^{-1}). These toxicity levels remained stable for one week and then began to rise quickly, reaching toxicities higher than 1400 μg STX eq 100 g^{-1} (1869, 2436, 1591, and 1442 μg STX eq 100 g^{-1} for Chanquín, Palihue, Deñal, and Rahue, respectively). On April 24, the toxicity increased again to reach values between 3044 and 6614 μg STX eq 100 g^{-1}. This increase was followed by the first massive beaching, which covered most Cucao beach (5 km long). At the end of April, the toxicity increased again to reach values near 5500 μg STX eq 100 g^{-1} in all localities and a second massive beaching was recorded. These toxicity levels remained approximately stable for 10 days from May 1 to May 11, a period during which at least three new surf clam beaching events were detected. The peak of shellfish toxicity was observed at May 20th, with values reaching 9059 µg STX eq 100 g^{-1} (6404, 7692, 8026, and 9059 µg STX eq 100 g^{-1} for Chanquín, Palihue, Deñal, and Rahue, respectively (Figures 2 and 3).

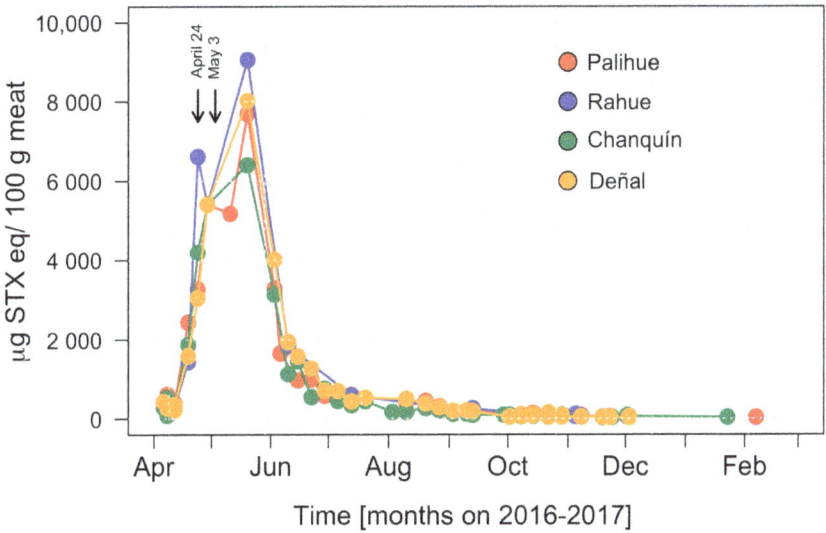

Figure 2. Toxicity changes in surf clams obtained in different locations from Cucao Bay during the March 2016–January 2017 toxic outbreak. Arrows indicates massive beaching on April 24 and May 3, 2016.

Since June 3, the toxicity decreased quickly to levels between 3137 and 4014 µg STX eq 100 g^{-1}. One week later, the toxicity additionally decreased to values below 2000 µg STX eq 100 g^{-1}. Since then, the toxicity declined gradually to 500 µg STX eq 100 g^{-1} by the end of July, and to 80 µg STX eq 100 g^{-1} (the regulatory limit) by October 29. On January 23, 2017 PSP was only detected in Chanquín (33 µg STX eq 100 g^{-1}).

2.2. Visual Observations of Surf Clams during Massive Beaching

On April 24, a seafood inspector of Laboratorio de Salud Pública Ambiental detected the first massive beaching of *M. donacium* covering most of Cucao beach. During this episode, thousands of individuals were found dead, dying, or paralyzed, lying on the sand surface. Two days later, the inspectors reported a notable decrease in the beaching surf clams, estimating that only 20% of individuals were lying on the sand. On May 3, during a new beaching episode, the density of surf clams on the sand surface was 40–50 individuals m^{-2}. Most of the specimens lying on the surface were alive with the foot or siphons extended or with the valve partially closed. A detailed examination revealed that the individuals were paralyzed because they had a weak reaction to mechanical stimulus on foot or siphons, with a slow and incomplete retraction. Three days later, the beaching surf clams decreased again, suggesting that some of the alive individuals recover the ability to re-burrow in the sand.

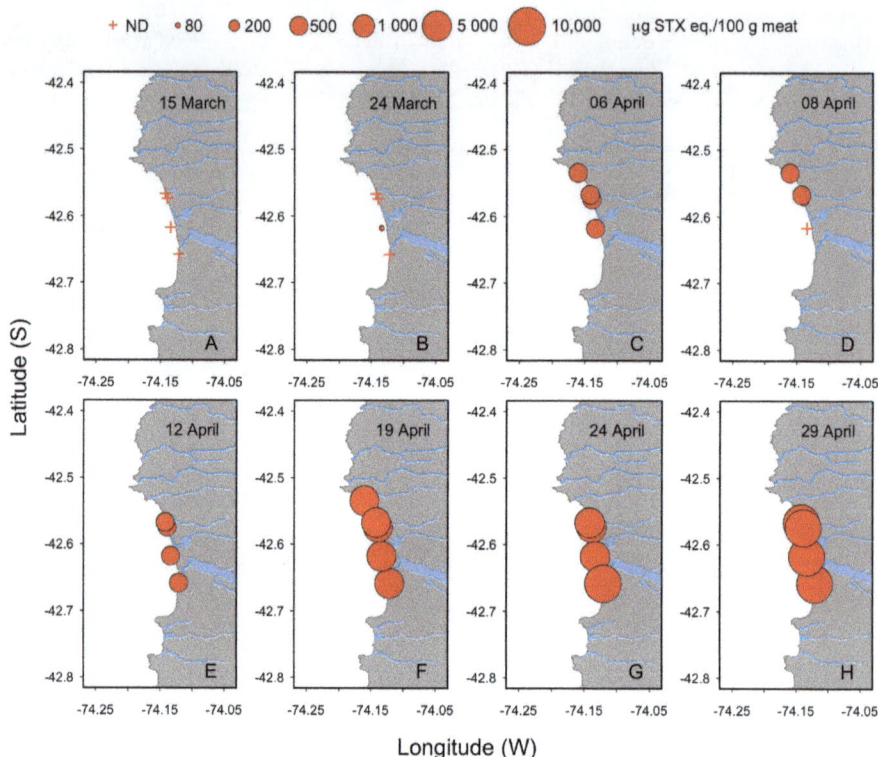

Figure 3. Temporal and spatial distribution of paralytic shellfish poisoning (PSP) toxicity in *M. donacium* along the coast from Cucao Bay during toxification phase of shellfish. (**A**) March 15th, (**B**) March 24th, (**C**) April 6th, (**D**) April, 8th, (**E**) April 12th, (**F**) April 19th, (**G**) April 24th and (**H**) April 29th.

2.3. Toxicity and Toxin Profile of Whole Individuals

The analyses of individual surf clams obtained in Chanquín on May 3, 2016 revealed a high toxicity with values between 1008 to 8763 µg STX eq 100 g^{-1} (4699 ± 2971 µg STX eq 100 g^{-1}), showing a bimodal distribution (Figure 4) with no individual with toxicities between 2000 and 4000 µg STX eq 100 g^{-1}. The toxin profile (% mole), was dominated by carbamoyl toxins that, in decreasing order of concentration, were GTX3 (24.64%), GTX1 (18.88%), GTX2 (14.79%), GTX4 (13.70%), neoSTX (7.46%), and STX (1.73%). Other toxins detected in less amount were N-sulfocarbamoyl, such as C1 (10.32%), C3 (5.89%), C2 (1%), and C4 (0.06%), and decarbamoyl, dcSTX being (1.17%) the most abundant toxins of this group, with dcGTX2 and dcGTX3 present in trace levels (<1%) (Figure 5).

2.4. Toxicity and Toxin Profile of Surf Clam Tissues

The most toxic organ was the digestive gland (DG), with 2221 ± 2193 µg STX eq 100 g^{-1}. On a molar basis, the toxin profile was dominated by carbamoyl toxins, GTX3 being (32.1%) the most abundant, followed by GTX2 (16.2%), GTX1 (16.1%), GTX4 (13.3%), neoSTX (5.9%), and STX (2.74%). The second group in importance was that of N-sulfocarbamoyl toxins, with C1, C3, C2, C4, and GTX5, in decreasing order of abundance. The less abundant of the analyzed toxins were dcGTX3 and dcGTX2 (Figure 6, Figure S1B).

The toxicity of all other body fractions (OBF) was the second in importance being approximately 1/3 as toxic as DG (710 ± 450 µg STX eq 100 g^{-1}). The toxin profile was dominated mainly by carbamoyl toxins (GTX3 46.5%, GTX4 16.4%, GTX2 13.3%, GTX1 9.5%, neoSTX 7.8%, and STX 3.8%).

In this case, the contribution of N-sulfocarbamoyl and decarbamoyl toxins were less relevant to the toxin profile (Figure 6).

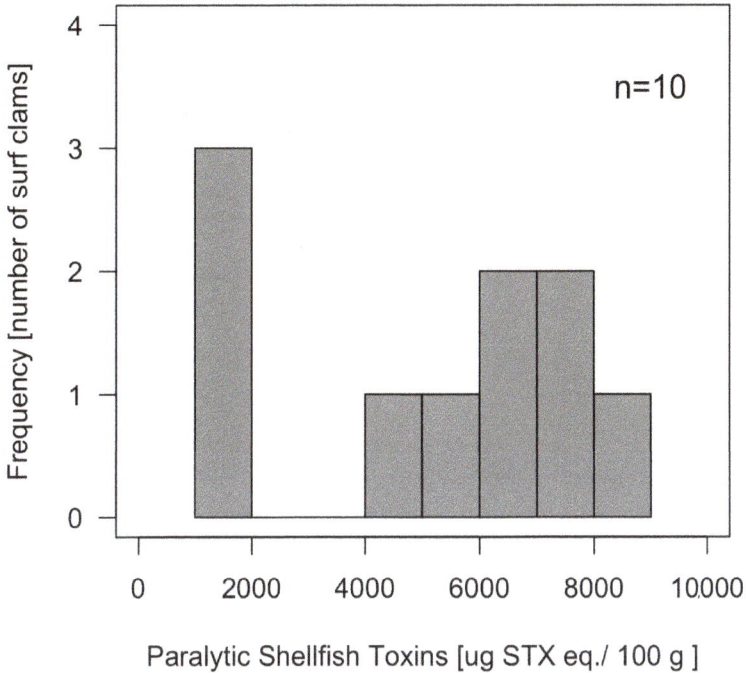

Figure 4. Interindividual toxicity of *M. donacium* obtained from Cucao Bay on 3 May 2016.

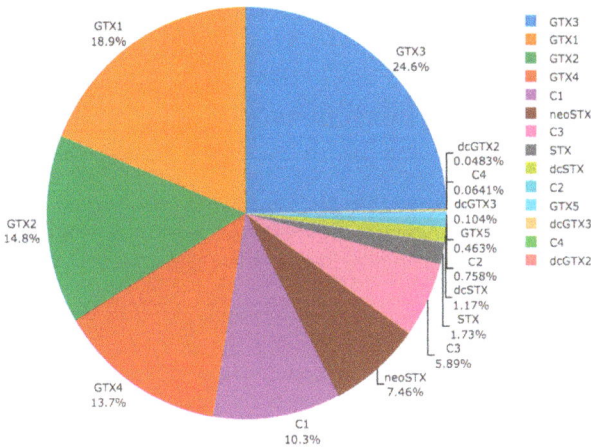

Figure 5. Relative toxin profile (% mole) of whole individuals of *Mesodesma donacium* collected from Cucao Bay (*n* = 10).

Foot (FT) had about 1/8 the toxicity of DG, with values of 297 ± 157 µg STX eq 100 g^{-1}. In this tissue, the toxin profile was dominated by carbamoyl toxins (GTX3 33.8%, GTX4 13.1%, GTX1 8%, GTX2 7.4%, neoSTX 4.6%, and STX 2.8%). Another relevant group was that of N-sulfocarbamoyl toxins, such as GTX5 (11.3%), C1 (6.9%), and C3 (6%). The toxins of the decarbamoyl group dcGTX2–3 were present in a very low molar percentage (Figure 6, Figure S1C).

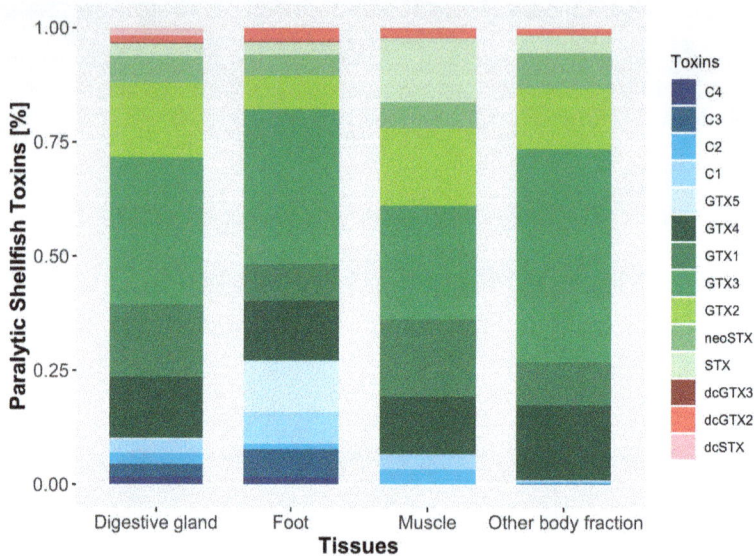

Figure 6. Relative toxin profile (% mole) of different organs of *Mesodesma donacium* collected from Cucao Bay (*n* = 10).

Finally, muscle (MU) had a toxicity of 314 ± 8.5 µg STX eq 100 g^{-1}. Its toxin composition (in molar proportion) was dominated by carbamoyl toxins (GTX3 24.8%, GTX2 17.1%, GTX1 16.9%, GTX4 12.6%, STX 14%, and neoSTX 5.6%), followed by other toxins present in low molar percentages as C1, C2, and dcGTX2 (Figure 6, Figure S1D).

2.5. Surf Clam Histology

Some pathological conditions were observed in *M. donacium* tissues exposed to the HAB of *A. catenella*. Bacterial colonies were found in the foot (Figure 7A), mantle, and muscle (with bacillary form in foot and mantle) in 40% of individuals. Additionally, cysts of digeneans metacercaria in the siphon muscle (Figure 7C, D) were found in all the analyzed individuals. Hemocyte aggregation in gill filaments was detected in 50% of individuals (Figure 7E) and haemocytic infiltration of the connective tissue surrounding the digestive gland tubules (Figure 7F) in 40%. No significant pathological changes were found in the foot (Figure 7B).

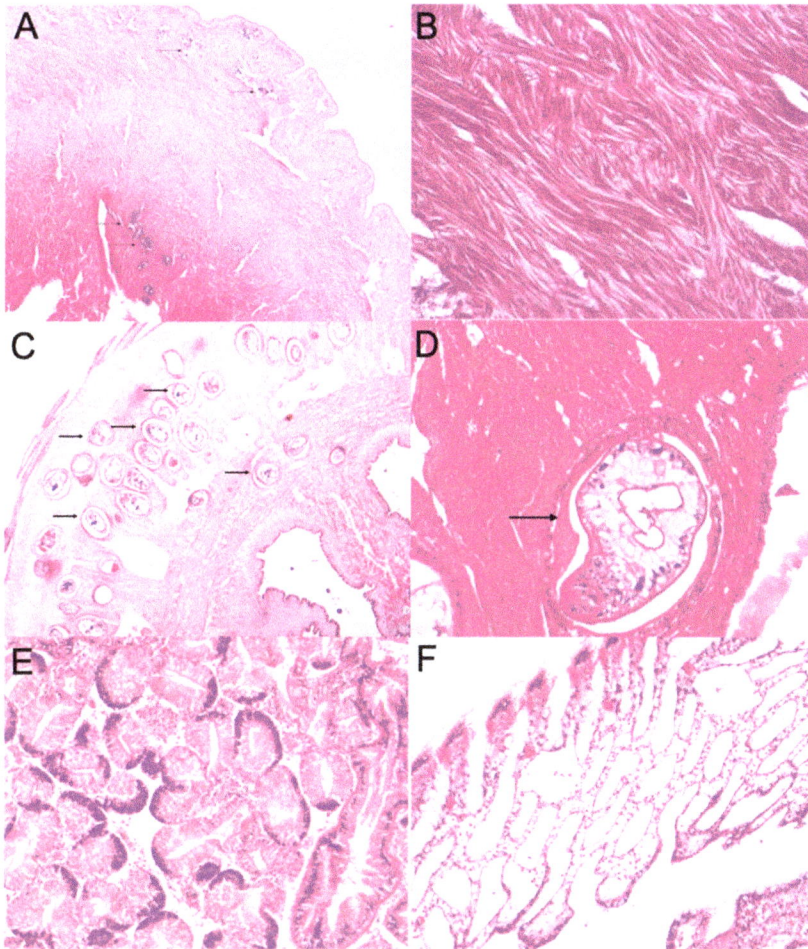

Figure 7. Histophatological section of *M. donacium* (H&E, medium magnification): (**A–B**) Foot; (**C–D**) Siphon; (**E**) Gill; and (**F**) Digestive gland. Cysts of digeneans metacercaria (arrows).

3. Discussion

During the bloom, *M. donacium* quickly bioaccumulated PSP toxins up to a maximum toxicity of 9059 µg STX eq 100 g^{-1}, which is 113 times the level considered hazardous for human consumption. This was the highest level ever recorded in this species since the monitoring program begun in 1995, attesting the magnitude and persistence of the toxic *A. catenella* bloom. The toxicity detected in *M. donacium* is higher than that found in other clams from southern Chile, such as *Venus antiqua* (111 µg STX eq 100 g^{-1}) and *Tagelus dombeii* (262 µg STX eq 100 g^{-1}), and similar to those reported in *Gari solida* (3286 µg STX eq 100 g^{-1}) [47]. However, the toxicity was lower than that reported in the Chilean blue mussels *Mytilus chilensis* 22,000 µg STX eq 100 g^{-1} [48], ribbed mussels *Aulacomya atra* in 1996 (113,259 µg STX eq 100 g^{-1}) [49], and recently, in 2018, when a global record of 143,000 µg STX eq 100 g^{-1} in *M. chilensis* [50] was attained.

Toxicity of the clams obtained during the third massive beaching ranged from 1008 to 8763 µg STX eq 100 g^{-1} (average 4699 µg STX eq 100 g^{-1}), revealing a high interindividual variability (CV = 67%) in PST bioaccumulation. This interindividual variation is similar to that reported in other shellfish, such as the clams *Arctica islandica* (56%) [51], *Spisula solidissima* (49%), and the scallop *Placopecten magellanicus* (44%) [51,52], but is lower than those reported in the clam *Panopea abrupta* (93%) [53]. The interindividual variability could be explained by diverse physiological processes, that, in turn, could be affected by the sensitivity of shellfish to PST toxins [54]. Diverse studies compared feeding behavior and physiology of different bivalve species, showing that the feeding response was correlated with the animal sensitivity to toxins and to the algal toxicity. For example, the clams *Mya arenaria*, *Ruditapes philippinarum*, and *Tagelus dombeii* and the oysters *Crassostrea virginica* and *Magallana gigas* (*Crassostrea gigas*) decrease the clearance rate on the presence of the toxic dinoflagellate *Alexandrium tamarense* [15,55–59]. The causes of the variability of toxin accumulation at interindividual level, notwithstanding, have been much less studied, but studies developed in *M. gigas* suggest that the feeding behavior is mainly responsible for interindividual variation in toxin bioaccumulation [54,60,61].

In the whole tissues of *M. donacium*, the toxin profile (in decreasing order) was dominated by GTX3, GTX1, GTX2, GTX4, C1, neoSTX, and C3, which contribute 95% of the total toxin content, while the remaining 5% corresponded to STX, dcSTX, C2, GTX5, C4, dcGTX2, and dcGTX3. The toxin profile is similar to those reported in the ribbed mussel *Aulacomya atra* from Darwin Channel, southern Chiloé (but differ in its higher proportion of GTX5) [49] and *M. chilensis* from Errazuriz Channel, Magallanes region, dominated mainly by GTX2 [62]. Similar profiles were found in the clams *Venus antiqua* and *Tagelus dombeii* (dominated by gonyautoxins, neoSTX and STX), but differ by the absence of C toxins [47]. Unfortunately, there is no information related to the toxic profile of *Alexandrium catenella* from Cucao Bay. The profile of *M. donacium*, however, showed characteristics which are very similar to that of strains of *A. catenella* isolated in other episodes, as the clones PFB36 from San Pedro Island (Los Lagos) and ACC02 from Coastal Channel (Aysén region). In both cases, the profile mainly differs by the presence of GTX6 in the dinoflagellates [33,63].

In the beached *M. donacium*, all tissues had a toxicity higher than the regulatory limit (80 µg STX eq 100 g^{-1}). In this shellfish, the relative contribution of the different tissues to toxicity (expressed as % µg STX eq) had the following pattern: digestive gland (DG) (68.4%), foot (FT) (23.7%), other body fractions (OBF) (7.4%), and MU (0.4%). In general, the anatomical distribution of toxicity in *M. donacium* is similar to those reported in other shellfish, in which DG is the initial repository of PST toxin during the intoxication phase [57]. Some examples of PST preferential accumulation in DG are the clams *Mya arenaria* (89.3%) [57], *Spisula solidissima* (85.6%) [64], *Mercenaria mercenaria* (80.9%) [65], and *Saxidomus giganteus* (66.2%) [57]. The FT tissues of *M. donacium* had a high contribution to toxicity (23.71%) as compared with the clam species mentioned above, in which this tissue only contributes 1.4 to 2.2% [57]. All tissues studied have a toxin profile similar to that of whole individuals in which the profile was dominated by GTX3, GTX2, GTX4, GTX1, neoSTX, and STX that contribute 97.3, 91.1, 86.2, and 69.6% for OBF, MU, DG, and FT, respectively. The most remarkable differences in toxin profile were observed in FT tissues with high contribution of GTX5 (11.3%), C1 (6.7%), C3 (6.0%), and dcGTX2 (3%). In this tissue, the high percentage of GTX5 suggests that a possible transformation of these toxin to STX by hydrolysis could be slower in comparison to the other surf clam tissues that have a low proportion of GTX5 (e.g. DG and OBF). Conversion of GTX5 to STX by hydrolysis was described in few bivalves as the mussel *Mytilus edulis* [66] and the clam *Paratapes undulatus* [67]. The high percentage α-isomers (C1, C3 in all tissues, and GTX1 in DG and MU) suggest other plausible transformation in *M. donacium* that corresponds to epimerization. In this case, the β-epimers (C2, C4, GTX3, and GTX4) will gradually convert into their thermodynamically more stable forms that correspond to α-isomers (C1, C3, GTX2, and GTX1) [6]. Epimerization is the most common bioconversion found in the bivalve tissues and has been described in the scallop *Pecten novaezelandiae* [68], mussels *M. edulis* [69] *M. galloprovincialis* [70,71], and the clam *Panopea globosa* [72]. Finally, the presence of dcGTX2, dcGTX3 (except in MU), and dcSTX (except in FT and MU) suggests the capability of *M. donacium* tissues to transform carbamoyl toxins GTX2, GTX3, and STX to their corresponding decarbamoyl derivatives and was demonstrated in clams *Protothaca staminea* [73], *Spisula solida* [74], *Mactra chinensis*, and *Perodina venulosa* [6].

Exposure to *A. catenella* seems to have induced immunological and inflammatory responses in *M. donacium*. Histopathological analyses in different tissues of the individuals exposed to the *A. catenella* bloom revealed the presence of hemocyte aggregation in gill tissues of 50% of analyzed individuals, that could be associated with an immunological response to *A. catenella* cells during filtration processes. In bivalves, the gill is one of the most important organs because of it is involved in respiration and feeding, being the first organ that has contact with toxic phytoplankton cells [75]. As in *M. donacium*, Estrada et al. [76] observed hemocyte aggregation in gill of the scallop, *Nodipecten subnodosus* exposed to the dinoflagellate PSP-producing *Gymnodinium catenatum*.

Inflammatory responses, as the observed haemocytic infiltration in connective tissues of digestive gland or other similar ones as diapedesis of hemocytes, were described in *Mytilus edulis*, *M. gigas*, and *Nodipecten subnodosus* exposed to *A. fudyense*, *A. minutum*, and *G. catenatum* [75–77] (all of them being PSP producers). Haberkorn et al. [75] proposed three hypotheses to explain inflammatory responses in this organ. The first suggested that the massive migration of hemocytes into the lumina of the digestive gland is a defense response of bivalves to protect tissues from toxicity or to remove toxic cells [77,78]. The second hypothesis proposed that hemocyte migration is a response to opportunistic bacterial infections that could appear by the exposure to toxic algae [77,78]. The third hypothesis suggested that the hemocyte diapedesis across intestine epithelia could be considered as a detoxification pathway [79].

In some individuals of beached surf clams, bacterial colonies were detected in the muscle, foot, and mantle and could be attributed to opportunistic bacteria of vibrio type. Abi-Khalil et al. [80] demonstrated that exposure to the neurotoxic *Alexandrium catenella* increases the susceptibility of *M. gigas* oysters to the pathogenic *Vibrio tasmaniensis* LGP32. In addition, they suggested that PSP toxins alone are not sufficient to induce mortalities but could rather participate in the induction of oyster mortality in a multifactorial way that involves vibrios and likely another pathogen.

The presence of cyst of digeneans metacercaria in siphon muscle is common in *M. donacium*, so it is not possible to infer if the *A. catenella* bloom had intensified the severity of infection or the impact on the bivalve population. López et al. [81] described a high prevalence of species belonging *Monorchiidae* family in bivalves obtained from different sites of Los Lagos region, including Cucao Bay. Lassudrie et al. [24], notwithstanding, suggested that the oyster *M. gigas* was more susceptible to trematode infestations when exposed *Alexandrium* blooms, which can produce degeneration of muscle fiber that could compromise valve-closure movements and the contraction of the adductor muscle.

During the bloom of *A. catenella*, a large number of individuals of *M. donacium* were either dead, dying, or paralyzed, lying on the sand surface. The observed reduction in foot contraction and valve gape and siphon retraction suggests a loss of nervous control due to PSP toxins. In *M. donacium*, the ability to re-burrow of some individuals from Cucao Bay, together with high interindividual variability (8-fold) and the bimodal distribution of toxin concentration, suggest the presence of two groups of individuals, one of them sensitive to the toxins (incapable of burrowing and with low toxin accumulation capability), and another one (more) resistant to the toxins (capable of burrowing and with high toxin accumulation capability), as described in *M. arenaria* [82,83]. PSP toxins can adversely affect susceptible marine invertebrates by blocking the Na^{++} influx in excitable cells and thus inhibiting the nerve action potential, leading to paralysis and death [84]. Indeed, the burrowing incapacitation produced by PSTs was described in the clam *Mya arenaria*, in which differences were found between resistant individuals, from populations frequently exposed to toxic *Alexandrium* blooms (Bay of Fundy, Canada), and sensitive individuals from areas with no record of PSP toxic episodes (St Lawrence Estuary, Canada) [15,85]. Considering the available information of the monitoring program and taking into account that this episode was the first detected along the open coast of the Pacific Ocean in southern Chiloé, it is very likely that the *M. donacium* population from Cucao Bay has not had a recurrent exposition to *A. catenella* and consequently that it has not been subjected to high selective pressure for PSP resistance. MacQuarrie and Bricelj [82] found high differences in toxicity accumulation in a laboratory experiment with *M. arenaria*, reporting a maximum toxicity in resistant individuals (77,000 µg STX eq 100 g^{-1}) 10-fold higher than the toxicity in sensitive ones (8200 µg STX eq 100 g^{-1}).

Our observations in this study suggest that the surf clam mortality during *A. catenella* toxic bloom in Cucao Bay was mainly due to incapability of burrowing—an indirect effect of PSP toxin-induced paralysis—which caused desiccation. However, more research is needed to determine the effects of PSP toxins on behavioral and physiological responses, nerve sensitivity, and genetic/molecular basis for the resistance or sensitivity of *M. donacium*.

4. Materials and Methods

4.1. Study Area

Cucao Bay (42°38′ S–74°07′ W) is located in the oceanic coast of the Chiloé Island (Figure 8). This site is characterized by the presence of large natural bed of surf clams (*M. donacium*) along a 20 km sandy beach. In general, this system is characterized by strong oceanic water influence, with salinity >31 despite the freshwater inputs from the Cucao River and other smaller tributaries (Figure 2). Water temperature ranges from 10 to 14 °C (IFOP, unpublished data) and the area is subjected to semidiurnal tides with amplitudes ranging from 2 m (neap tides) to 4 m (spring tides) (www.shoa.cl).

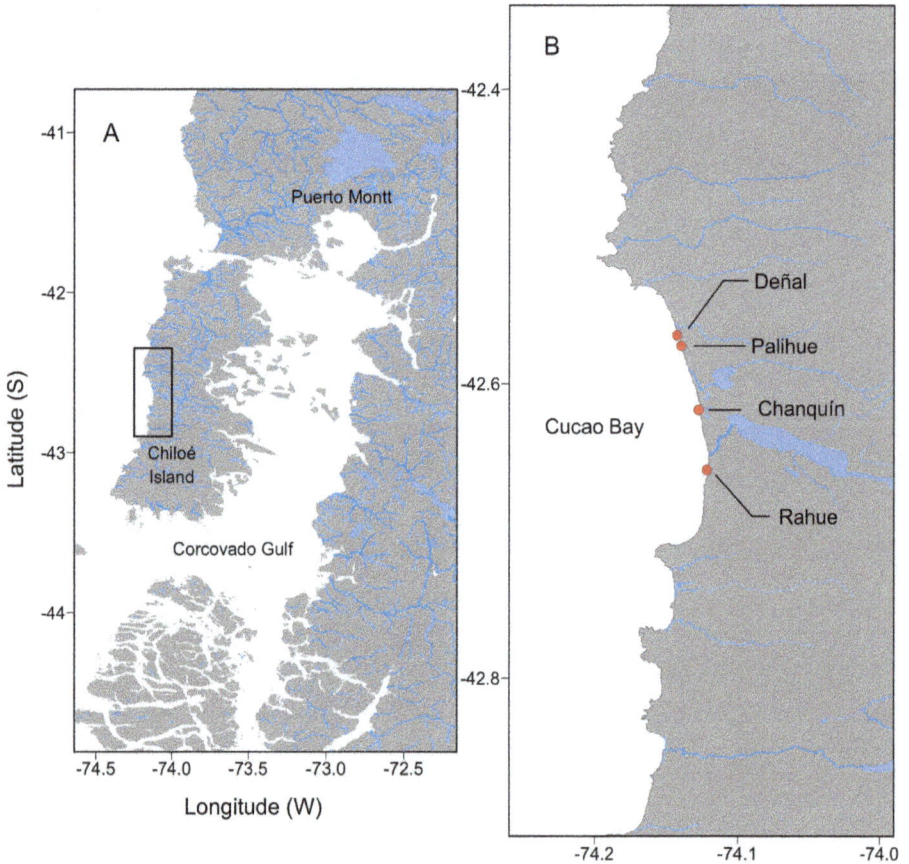

Figure 8. Study area in southern Chiloé showing (**A**) The map shows a section of the Chilean Inland Sea; (**B**) The four sampling stations at Cucao Bay in the oceanic coast of the Chiloé Island.

4.2. Shellfish Sampling, Toxin Extraction, and Biological Method Analyses Used for Rutine Monitoring

Within the framework of the regional monitoring program developed by the Laboratorio de Salud Pública Ambiental (Secretaría Regional Ministerial de Salud, Region de Los Lagos), surf clam samples were obtained periodically from Cucao Bay at least twice a month—from March 15, 2016, to January 23, 2017—in four sampling stations located at Cucao Bay: Deñal, Palihue, Chanquín, and Rahue (Figure 8).

Each sample consisted of at least 20 individuals of commercial size (>6 cm) collected by fishermen by means of hang-gathering. Samples were placed inside a plastic bag, then placed in a cool box at 10 °C and transported to the laboratory. Upon arrival to the laboratory, paralytic shellfish toxins (PST) were extracted from surf clam tissues following the official AOAC method 959.08 [86]. For toxin extraction, 100 g of homogenized raw tissues were mixed with 100 mL of HCl (0.1 N) solution using a blender and then boiled for 5 min. The sample was cooled at room temperature for 10 min, then the pH was corrected to between 2–4. At the end of the procedure, the resulting extract was transferred to a 200 mL volumetric flask and filled up to the mark with HCl (0.003 N). Aliquots (1 mL) of the final extract were intraperitoneally injected into three Swiss mice weighing 19–21 g following the official AOAC method 959.08, and their death times were recorded. If any mouse died in less than 5 min, the test was performed again using diluted samples until the mouse died within 5 to 7 min. The toxicity was calculated and expressed as μg STX eq 100 g^{-1} sample using Sommer's Table.

4.3. Shellfish Sampling, Toxin Extraction, Chromatographic, and Histopathological Analyses during Beaching

Data: Surf clam samples were collected from Bahía Cucao during a massive beaching detected on May 3, 2016. The sample consisted of 20 individuals of commercial size (>6 cm) collected from the surface of the sand. To ensure that the shellfish were still alive, a mechanical stimulus was induced by means of a puncture in the foot using a steel needle to determine foot contraction, incapability, or reduction of valve closure and siphon retraction. Upon arrival to the laboratory, the shellfish were washed and distributed into two subsamples completely at random. The first subsample consisted of 10 clams and was used for individual whole sample analyses. The second subsample contained 10 individuals and was used to determine the anatomical distribution. These individuals were carefully dissected into foot (FT, the edible tissue) and the nonedible tissues corresponding to digestive gland (DG), adductor muscle (MU), and all other body fractions (OBF) (including mantle, gill, kidney, and siphons). Each tissue was extracted following the official AOAC method 2011.02 [87], with slight modifications. Extraction was performed with HCl (0.1 N) (1:1, *w/v*) using an Ultra-Turrax T25 dispersing system (IKA®Werke GmbH & Co. KG, Staufen, Germany) at 11,000 rpm for 3 min. After each extraction, the pH of the resulting emulsion was adjusted to 2–4, boiled at 100 °C for 5 min, centrifuged at 5000 *g* for 20 min (Centurion K2015R, Centurion Scientific Ltd, Stoughton, West Sussex, UK), and its pH adjusted again to 2–4. To deproteinate the samples, a 1 mL aliquot was mixed with 50 μL of 30% trichloroacetic acid (TCA), vortex mixed, and then centrifuged at 9000 *g* for 5 min. To neutralize the solution, 70 μL of NaOH (1 M) was added, and then vortexed and centrifuged at 9000 *g* for 5 min. One 500-μL aliquot was filtered through a 0.2 μm Clarinert nylon syringe filter (13 mm diameter) (Agela technologies, Torrence, CA, USA) and stored into an autosampler vial. A second 500-μL aliquot was hydrolyzed in order to transform the sulfocarbamate toxins (if present) to the corresponding carbamate toxins (HCl 0.4 N, 100 °C, 15 min), filtered through 0.2 μm nylon filter, and stored into an autosampler vial. For some analyses, extracts were diluted 5-fold or 10-fold with HCl 0.1 N to ensure the correct quantification of the most abundant toxins in the samples.

The most of important PST toxins were quantified following the official AOAC method 2011.02 by high performance liquid chromatography (HPLC) using a Hitachi LaChrom Elite HPLC system equipped with a Hitachi FL detector L2485 (Hitachi High Technologies America Inc., Chatsworth, CA, USA) and a post-column reaction system composed by two Waters pumps (Water Reagent Manager) and a 20 m Teflon coil (2 mL) in a heating reactor (Pickering Laboratories, Mountain View, CA, USA). The chromatographic separation was carried out using a Zorbax Bonus RP column (150 × 4.6 mm, 3.5 μm particle diameter) (Agilent, Santa Clara, CA, USA) with a Zorbax Bonus RP guard column

(12.5 × 4.6 mm, 5 μm particle diameter) at 40 °C. For the separation, two mobile phases were used consisting of 11 mM heptanesulfonate and 5.5 mM phosphoric acid solution adjusted to pH 7.1 with ammonium hydroxide (A) and 11 mM heptane sulfonate, 16.5 mM phosphoric acid, 11.5% acetonitrile solution adjusted to pH 7.1 with ammonium hydroxide (B). A gradient elution started with a proportion of 100% A, which was maintained for 8 minutes, followed by a linear increase to 100% B from minute 8.01 to minute 15, and then held for 4 min. The column was equilibrated in the initial conditions for 5 min previously to next run. The mobile phase flow was 0.8 mL min^{-1} and the injection volume 10 μL. After separation, the toxins were derivatized in the post-column reaction system (85 °C) with a solution of 100 mM phosphoric acid, 5 mM periodic acid solution adjusted to pH 7.8 with 5 M sodium hydroxide, and, at the end of the reaction coil, mixed with a solution of 0.75 M nitric acid. The flow rate of both solutions was 0.4 mL min^{-1}. Finally, the detection of the fluorescent derivatives of the toxins was carried out with an FL detector set to 330/390 nm excitation/emission wavelengths. PST concentration in the samples was quantified by comparing the obtained response with that of corresponding reference materials (from IMB-NRC, Ottawa, ON, Canada). The limits of detection of the technique were estimated for a signal/noise ratio of 3 (Table S1). Finally, to determine the total toxicity for each sample, equivalency factors (TEFs) were used for calculations [88].

Upon arrival to the laboratory, 10 individuals were carefully dissected for histopathological studies in siphons (SI), gill (GI), gonad (GO), digestive gland (DG), foot (FT), and mantle (MA). Each tissue sample was placed in a cassette, immediately fixed in Davidson´s solution [89] and kept at 4 °C for 48 h before being switched to ethanol (70%). The tissue samples were then dehydrated in ascending ethanol solutions, cleared with xylene, and embedded in paraffin wax using a spin tissue processor (STP 120, Microm International GmBH Thermo Scientific, Walldorf, Germany). Finally, 5 μm thick sections of each tissue were obtained using a rotary microtome (Microm HM325, Microm International GmBH Thermo Scientific, Walldorf, Germany), mounted on slides and stained with Harry´s hematoxylin and eosin [90]. All sample tissues were examined under a light microscope (Leica DFC295) and micrographs of the material were obtained by means a microscope camera (Leica DM 2000 LED).

Supplementary Materials: The following are available online at http://www.mdpi.com/2072-6651/11/4/188/s1, Figure S1: Selected chromatograms of Paralytic Shellfish Toxins (PST) in reference material (**A**), digestive gland (**B**), foot (**C**) and muscle (**D**), Table S1: Limits of detection (LOD) and quantification (LOQ) for each toxin in *Mesodesma donacium*.

Author Contributions: Conceptualization: G.Á., P.A.D., M.G., E.U. and J.B.; Methodology: G.Á., M.G., C.H. and M.A.; Formal Analysis G.Á., P.A.D., M.G., M.A., E.U. and J.B.; Investigation: G.Á., P.A.D., M.G., M.A., I.G., R.P., F.Á., J.R., C.H., E.U., J.B.; Writing—Original Draft Preparation: G.Á., P.A.D., M.G. and J.B.; Writing—Review & Editing: G.Á., P.A.D., M.G. and J.B.; Visualization: G.Á., P.A.D. and J.B.; Funding acquisition: G.Á., M.G. and P.A.D.

Funding: Gonzalo Álvarez and Patricio A. Díaz were funded by the Chilean National Commission for Scientific and Technological Research (CONICYT + PAI/CONCURSO NACIONAL INSERCION EN LA ACADEMIA CONVOCATORIA 2015, 79150008 (G. Álvarez), CONVOCATORIA 2016, 79160065 (P.A. Díaz)).

Acknowledgments: We would like to thank to Pedro Toledo Director of Aquaculture Department of Universidad Católica del Norte for his support. We thank CYTED Program (AquaCibus network 318RT0549 "Strengthening aquaculture in Iberoamérica: quality, competitiveness and sustainability") for promoting interactions among authors.

Conflicts of Interest: The authors declare no conflict of interest.

References

1. Mons, M.N.; Van Egmond, H.P.; Speijers, G.J.A. Paralytic shellfish poisoning: A review. *Rivm Rep.* **1998**, *388802005*, 1–47.
2. Anderson, D.M.; Alpermann, T.J.; Cembella, A.; Collos, Y.; Masseret, E.; Montresor, M. The globally distributed genus *Alexandrium*: Multifaceted roles in marine ecosystems and impacts on human health. *Harmful Algae* **2012**, *14*, 10–35. [CrossRef]

3.   Anderson, D.M. HABs in a changing world: A perspective on harmful algal blooms, their impacts, and research and management in a dynamic era of climactic and environmental change. In *Harmful Algae 2012, Proceedings of the 15th International Conference on Harmful Algae, International Society for the Study of Harmful Algae (2014). ISBN 978-87-990827-4-2, 16p*; Kim, H.G., Reguera, B., Hallegraeff, G., Lee, C.K., Han, M.S., Choi, J.K., Eds.; 2014. Available online: https://www.ncbi.nlm.nih.gov/pmc/articles/PMC4667985/ (accessed on 2 December 2015).

4.   Usup, G.; Ahmad, A.; Matsuoka, K.; Lim, P.T.; Leaw, C.P. Biology, ecology and bloom dynamics of the toxic marine dinoflagellate *Pyrodinium bahamense*. *Harmful Algae* **2012**, *14*, 301–312. [CrossRef]

5.   Wiese, M.; D'Agostino, P.M.; Mihali, T.K.; Moffitt, M.C.; Neilan, B.A. Neurotoxic Alkaloids: Saxitoxin and Its Analogs. *Mar. Drugs* **2010**, *8*, 2185–2211. [CrossRef] [PubMed]

6.   Oshima, Y. Chemical and enzymatic transformation of paralytic shellfish toxins in marine organisms. In *Harmful Marine Algal Blooms*; Lassus, P., Arzul, G., Erard-Le Denn, E., Gentien, P., Marcaillou-Le Baut, C., Eds.; Lavoisier Science Publishers: Paris, France, 1995; pp. 475–480.

7.   Oshima, Y.; Sugino, K.; Itakura, H.; Hirota, M.; Yasumoto, Y. Comparative studies on paralytic shellfish toxin profile of dinoflagellates and bivalves. In *Toxic Marine Phytoplankton*; Granelli, E., Sundstrom, B., Edler, L., Anderson, D.M., Eds.; Elsevier Science Publishing: New York, NY, USA, 1990; pp. 391–396.

8.   Negri, A.; Stirling, D.; Quilliam, M.; Blackburn, S.; Bolch, C.; Burton, I.; Eaglesham, G.; Thomas, K.; Walter, J.; Willis, R. Three novel hydroxybenzoate saxitoxin analogues isolated from the dinoflagellate *Gymnodinium catenatum*. *Chem. Res. Toxicol.* **2003**, *16*, 1029–1033. [CrossRef]

9.   Costa, P.R.; Robertson, A.; Quilliam, M.A. Toxin profile of *Gymnodinium catenatum* (Dinophyceae) from the Portuguese Coast, as determined by Liquid Chromatography Tandem Mass Spectrometry. *Mar. Drugs* **2015**, *13*, 2046–2062. [CrossRef]

10. Vale, P. New saxitoxin analogues in the marine environment: Developments in toxin chemistry, detection and biotransformation during the 2000s. *Phytochem. Rev.* **2010**, *9*, 525–535. [CrossRef]

11. Dell' Aversano, C.; Walter, J.A.; Burton, I.W.; Stirling, D.J.; Fattorusso, E.; Quilliam, M. Isolation and structure elucidation of new and unusual saxitoxin analogues from mussels. *J. Nat. Prod.* **2008**, *71*, 1518–1523. [CrossRef]

12. Landsberg, J.H. The effects of harmful algal blooms on aquatic organisms. *Rev. Fish. Sci.* **2002**, *10*, 113–390. [CrossRef]

13. Shumway, S.E. A review of the effects of algal blooms on shellfish and aquaculture. *J. World Aquac. Soc.* **1990**, *21*, 65–104. [CrossRef]

14. Shumway, S.E.; Cucci, T.L. The effects of the toxic dinoflagellate *Protogonyaulax tamarensis* on the feeding and behaviour of bivalve molluscs. *Aquat. Toxicol.* **1987**, *10*, 9–97. [CrossRef]

15. Bricelj, M.; Cembella, A.; Laby, D.; Shumway, S.; Cucci, T. Comparative physiological and behavioral responses to PSP toxins in two bivalve molluscs, the softshell Clam, *Mya arenaria*, and surfclam, *Spisula solidissima*. In *Harmful and Toxic Algal Blooms*; Yasumoto, T., Oshima, Y., Fukuyo, Y., Eds.; Intergovermental Oceanographic Comission of UNESCO: Paris, France, 1996; pp. 405–408.

16. Gainey, L.F., Jr.; Shumway, S.E. A compendium of the responses of bivalve molluscs to toxic dinoflagellates. *J. Shellfish Res.* **1988**, *7*, 623–628.

17. Lesser, M.P.; Shumway, S.E. Effects of toxic dinoflagellates on clearance rates and survival in juvenile bivalve molluscs. *J. Shellfish Res.* **1993**, *12*, 377–381.

18. Hegaret, H.; Wikfors, G.H.; Shumway, S.E. Diverse feeding responses of five species of bivalve mollusc when exposed to three species of harmful algae. *J. Shellfish Res.* **2007**, *26*, 549–559. [CrossRef]

19. Lassus, P.; Bardouil, M.; Beliaeff, B.; Masselin, P.; Naviner, M.; Truquet, P. Effect of a continuous supply of the toxic dinoflagellate *Alexandrium minutum* Halim on the feeding behavior of the Pacific oyster (*Crassostrea gigas* Thunberg). *J. Shellfish Res.* **1999**, *18*, 211–216.

20. May, S.P.; Burkholder, J.A.M.; Shumway, S.E.; Hégaret, H.; Wikfors, G.H.; Frank, D. Effects of the toxic dinoflagellate *Alexandrium monilatum* on survival, grazing and behavioral response of three ecologically important bivalve molluscs. *Harmful Algae* **2010**, *9*, 281–293. [CrossRef]

21. Tran, D.; Haberkorn, H.; Soudant, P.; Ciret, P.; Massabuau, J.-C. Behavioral responses of *Crassostrea gigas* exposed to the harmful algae *Alexandrium minutum*. *Aquaculture* **2010**, *298*, 338–345. [CrossRef]

22. Fabioux, C.; Sulistiyani, Y.; Haberkorn, H.; Hégaret, H.; Amzil, Z.; Soudant, P. Exposure to toxic *Alexandrium minutum* activates the detoxifying and antioxidant systems in gills of the oyster *Crassostrea gigas*. *Harmful Algae* **2015**, *48*, 55–62. [CrossRef]
23. Neves, R.A.F.F.; Figueiredo, G.M.; Valentin, J.L.; da Silva Scardua, P.M.; Hégaret, H. Immunological and physiological responses of the periwinkle *Littorina littorea* during and after exposure to the toxic dinoflagellate *Alexandrium minutum*. *Aquat. Toxicol.* **2015**, *160*, 96–105. [CrossRef] [PubMed]
24. Lassudrie, M.; Wikfors, G.H.; Sunila, I.; Alix, J.H.; Dixon, M.S.; Combot, D.; Soudant, P.; Fabioux, C.; Hégaret, H. Physiological and pathological changes in the eastern oyster *Crassostrea virginica* infested with the trematode *Bucephalus* sp. and exposed to the toxic dinoflagellate *Alexandrium fundyense*. *J. Invertebr. Pathol.* **2015**, *126*, 51–63. [CrossRef] [PubMed]
25. Borcier, E.; Morvezen, R.; Boudry, P.; Miner, P.; Charrier, G.; Laroche, J.; Hegaret, H. Effects of bioactive extracellular compounds and paralytic shellfish toxins produced by *Alexandrium minutum* on growth and behaviour of juvenile great scallops *Pecten maximus*. *Aquat. Toxicol.* **2017**, *184*, 142–154. [CrossRef] [PubMed]
26. Horstman, D.A. Reported red-water outbreaks and their effects on fauna of the west and south coasts of South Africa, 1959–1980. *Fish. Bull. South. Afr.* **1981**, *15*, 71–88.
27. Popkiss, M.E.E.; Horstman, D.A.; Harpur, D. Paralytic shellfish poisoning: A report of 17 cases in Cape Town. *South Afr. Med. J.* **1979**, *55*, 1017–1023.
28. Wardle, W.J.; Ray, S.M.; Aldrich, A.S. Mortality of marine organisms associated with offshore summer blooms of the toxic dinoflagellate *Gonyaulax monilata* Howell at Galveston, Texas. In *Proceedings of the First International Conference on Toxic Dinoflagellate Blooms*; LoCicero, V.R., Ed.; Massachusetts Science and Technology Foundation: Wakefield, MA, USA, 1975; pp. 257–263.
29. Guzmán, L.; Campodonico, I.; Antunovic, M. Estudios sobre un florecimiento toxico causado por *Gonyaulax catenella* en Magallanes. IV. Distribución y niveles de veneno paralítico de los mariscos (noviembre de 1972–noviembre de 1973). *Inst. Patagon.* **1975**, *6*, 209–223.
30. Lembeye, G.; Marcos, N.; Sfeir, A.; Molinet, C.; Jara, F. *Seguimiento de la toxicidad en recursos pesqueros de importancia comercial en la X y XI región*; Informe Final Proyecto FIP 97/49; Universidad Austral de Chile: Puerto Montt, Chile, 1998; Volume 89.
31. García, C.; Mardones, P.; Sfeir, A.; Lagos, N. Simultaneous presence of Paralytic and Diarrheic Shellfish Poisoning toxins in *Mytilus chilensis* samples collected in the Chiloe Island, Austral Chilean Fjords. *Biol. Res.* **2004**, *37*, 721–731. [CrossRef]
32. Muñoz, P.; Avaria, S.; Sievers, H.; Prado, R. Presencia de dinoflagelados toxicos del genero *Dinophysis* en el seno Aysén, Chile. *Rev. Biol. Mar.* **1992**, *27*, 187–212.
33. Aguilera-Belmonte, A.; Inostroza, I.; Franco, J.M.; Riobo, P.; Gómez, P.I. The growth, toxicity and genetic characterization of seven strains of *Alexandrium catenella* (Whedon and Kofoid) Balech 1985 (Dinophyceae) isolated during the 2009 summer outbreak in southern Chile. *Harmful Algae* **2011**, *12*, 105–112. [CrossRef]
34. Díaz, P.A.; Molinet, C.; Seguel, M.; Díaz, M.; Labra, G.; Figueroa, R. Coupling planktonic and benthic shifts during a bloom of *Alexandrium catenella* in southern Chile: Implications for bloom dynamics and recurrence. *Harmful Algae* **2014**, *40*, 9–22. [CrossRef]
35. Guzmán, L.; Pacheco, H.; Pizarro, G.; Alárcon, C. *Alexandrium catenella* y veneno paralizante de los mariscos en Chile. In *Floraciones Algales Nocivas en el Cono Sur Americano*; Sar, E.A., Ferrario, M.E., Reguera, B., Eds.; Instituto Español de Oceanografía: Madrid, Spain, 2002; Volume 11, pp. 235–255.
36. Mardones, J.; Clément, A.; Rojas, X.; Aparicio, C. *Alexandrium catenella* during 2009 in Chilean waters, and recent expansion to coastal ocean. *Harmful Algae News* **2010**, *41*, 8–9.
37. Molinet, C.; Lafón, A.; Lembeye, G.; Moreno, C.A. Patrones de distribución espacial y temporal de floraciones de *Alexandrium catenella* (Whedon & Kofoid) Balech 1985, en aguas interiores de la Patagonia noroccidental de Chile. *Rev. Chil. Hist. Nat.* **2003**, *76*, 681–698.
38. Díaz, P.A.; Molinet, C.; Seguel, M.; Díaz, M.; Labra, G.; Figueroa, R.I. Species diversity and abundance of dinoflagellate resting cysts seven months after a bloom of *Alexandrium catenella* in two contrasting coastal systems of the Chilean Inland Sea. *Eur. J. Phycol.* **2018**, *3*, 410–421. [CrossRef]
39. Guzmán, L.; Espinoza-González, O.; Pinilla, E.; Martinez, R.; Carbonell, P.; Calderón, M.J.; Lopez, L.; Hernández, C. The *Alexandrium catenella* and PSP outbreak in the Chilean coast, the first in the open coast of the South East Pacific Ocean. In *The 17th International Conference of Harmful Algae*; Odebrecht, C., Cintra, F., Proença, L., Mafra, L., Schramm, M., Garlet, N., Alves, T., Eds.; Florianopolis, Brazil, 2016; Volume 74.

40. Hernández, C.; Díaz, P.A.; Molinet, C.; Seguel, M. Exceptional climate anomalies and northwards expansion of Paralytic Shellfish Poisoning outbreaks in Southern Chile. *Harmful Algae News* **2016**, *54*, 1–2.

41. Alamo, V.V.; Valdivieso, V.M. *Lista sistemática de moluscos marinos del Perú*; Segunda edición, revisada y actualizada; Instituto del Mar del Perú: Callao, Perú, 1997; p. 183.

42. Tarifeño, E. Studies on the biology of surf clam *Mesodesma donacium* (Lamarck, 1818) (Bivalvia: Mesodesmatidae) from Chilean sandy beaches. Ph.D. Dissertation, University of California, Los Angeles, CA, USA, 1980.

43. Guzmán, N.; Saá, S.; Ortlieb, L. Catálogo descriptivo de los moluscos litorales (Gastropoda y Pelecypoda) de la zona de Antofagasta, 23 S (Chile). *Est. Ocea.* **1998**, *17*, 17–86.

44. Jaramillo, E.; Pino, M.; Filun, L.; Gonzalez, M. Longshore distribution of *Mesodesma donacium* (Bivalvia: Mesodes-matidae) on a sandy beach of the south of Chile. *Veliger* **1994**, *37*, 192–200.

45. Ortega, L.; Castilla, J.C.; Espino, M.; Yamashiro, C.; Defeo, O. Effects of fishing, market price, and climate on two South American clam species. *Mar. Ecol. Prog. Ser.* **2012**, *469*, 71–82. [CrossRef]

46. Riascos, J.M.; Carstensen, D.; Laudien, J.; Arntz, W.E.; Oliva, M.E.; Guntner, A.; Heilmayer, O. Thriving and declining: Climate variability shaping life-history and population persistence of *Mesodesma donacium* in the Humboldt Upwelling System. *Mar. Ecol. Prog. Ser.* **2009**, *385*, 151–163. [CrossRef]

47. García, C.; Pérez, F.; Contreras, C.; Figueroa, D.; Barriga, A.; López-Rivera, A.; Araneda, O.F.; Contreras, H.R. Saxitoxins and okadaic acid group: Accumulation and distribution in invertebrate marine vectors from Southern Chile. *Food Addit. Contam. Part A* **2015**, *32*, 984–1002. [CrossRef]

48. Molinet, C.; Niklitschek, E.; Seguel, M.; Díaz, P. Trends of natural accumulation and detoxification of paralytic shellfish poison in two bivalves from the Northwest Patagonian inland sea. *Revista de Biologia Marina y Oceanografía* **2010**, *45*, 195–204. [CrossRef]

49. Compagnon, D.; Lembeye, G.; Marcos, N.; Ruíz-Tagle, N.; Lagos, N. Accumulation of Paralytic Shellfish Poisoning toxins in the bivalve *Aulacomya ater* and two carnivorous gastropods *Concholepas concholepas* and *Argobuccinum ranelliformes* during an *Alexandrium catenella* bloom in southern Chile. *J. Shellfish Res.* **1998**, *17*, 67–73.

50. Díaz, P.A.; Álvarez, A.; Varela, D.; Pérez-Santos, I.; Díaz, M.; Molinet, C.; Seguel, M.; Aguilera-Belmonte, A.; Guzmán, L.; Uribe, E.; et al. Impacts of harmful algal blooms on the aquaculture industry: Chile as a case study. *Perspect. Phycol.* **2019**. [CrossRef]

51. White, A.W.; Shumway, S.E.; Nassif, J.; Whittaker, D.K. Variation in Levels of Paralytic Shellfish Toxins among Individual Shellfish. In *Toxin Phytoplankton Blooms in the Sea*; Smayda, T.J., Shimizu, Y., Eds.; Elsevier Science Publishers B.V.: Amsterdam, The Netherlands, 1993; pp. 441–446.

52. Cembella, A.D.; Shumway, S.E.; Lewis, N.I.I. Anatomical distribution and spatio-temporal variation in paralytic shellfish toxin composition in two bivalve species from the Gulf of Maine. *J. Shellfish Res.* **1993**, *12*, 389–403.

53. Curtis, K.M.; Trainer, V.L.; Shumway, S.E. Paralytic shellfish toxins in geoduck clams (Panope abrupta): Variability, anatomical distribution, and comparison of two toxin detection methods. *J. Shellfish Res.* **2000**, *19*, 313–319.

54. Pousse, E.; Flye-Sainte-Marie, J.; Alunno-Bruscia, M.; Hegaret, H.; Jean, F. Sources of paralytic shellfish toxin accumulation variability in the Pacific oyster *Crassostrea gigas*. *Toxicon* **2018**, *144*, 14–22. [CrossRef]

55. Mardsen, I.D.; Shumway, S.E. The effect of a toxic dinoflagellate (*Alexandrium tamarense*) on the oxygen uptake of juvenile filter-feeding bivalve molluscs. *Comp. Biochem. Physiol.* **1993**, *106*, 769–773.

56. Bardouil, M.; Bohec, M.; Cormerais, M.; Bougrier, S.; Lassus, P. Experimental study of the effects of a toxic microalgal diet on feeding of the oyster *Crassostrea gigas* Thunberg. *J. Shellfish Res.* **1993**, *12*, 417–422.

57. Bricelj, V.M.; Shumway, S. Paralytic shellfish toxins in bivalve molluscs: Occurrence, transfer kinetics, and biotransformation. *Rev. Fish. Sci.* **1998**, *6*, 315–383. [CrossRef]

58. Li, S.C.; Wang, W.X.; Hsieh, D.P.H. Effects of toxic dinoflagellate *Alexandrium tamarense* on the energy budgets and growth of two marine bivalves. *Mar. Environ. Res.* **2002**, *53*, 145–160. [CrossRef]

59. Navarro, J.M.; González, K.; Cisternas, B.A.; López, J.A.; Chaparro, O.R.; Segura, C.; Cordova, M.; Suarez Isla, B.; Fernández-Reiriz, M.J.; Labarta, U. Contrasting physiological responses of two populations of the razor clam *Tagelus dombeii* with different histories of exposure to paralytic shellfish poisoning (PSP). *PLoS ONE* **2014**, *9*, 9. [CrossRef]

60. Bougrier, S.; Lassus, P.; Bardouil, M.; Masselin, P.; Truquet, P. Paralytic shellfish poison accumulation yields and feeding time activity in the Pacific oyster (*Crassostrea gigas*) and king scallop (*Pecten maximus*). *Aquat. Living Resour.* **2003**, *16*, 347–352. [CrossRef]

61. Haberkorn, H.; Tran, D.; Massabuau, J.-C.; Ciret, P.; Savar, V.; Soudant, P. Relationship between valve activity, microalgae concentration in the water and toxin accumulation in the digestive gland of the Pacific oyster *Crassostrea gigas* exposed to *Alexandrium minutum*. *Mar. Pollut. Bull.* **2011**, *62*, 1191–1197. [CrossRef]

62. Lagos, N.; Compagnon, D.; Seguel, M.; Oshima, Y. Paralytic shellfish toxin composition: A quantitative analysis in Chilean mussels and dinoflagellate. In *Harmful Toxic Algal Blooms*; Yasumoto, T., Oshima, Y., Fukuyo, Y., Eds.; Intergovermental Oceanographic Comission of UNESCO: Paris, France, 1996; pp. 121–124.

63. Krock, B.; Seguel, C.; Cembella, A. Toxin profile of *Alexandrium catenella* from the Chilean coast as determined by liquid chromatography with fluorescence detection and liquid chromatography coupled with tandem mass spectrometry. *Harmful Algae* **2007**, *6*, 734–744. [CrossRef]

64. Bricelj, V.M.; Cembella, A.D. Fate of gonyautoxins in surfclams, *Spisula solidissima*, grazing upon toxigenic *Alexandrium*. In *Harmful Marine Algal Blooms*; Lassus, P., Arzul, G., Erard, E., Gentien, P., Marcaillou, C., Eds.; Lavoisier, Intercept Ltd: Paris, France, 1995; pp. 413–418.

65. Bricelj, V.M.; Lee, J.H.; Cembella, A.D. Influence of dinoflagellate cell toxicity on uptake and loss of paralytic shellfish toxins in the northern quahog *Mercenaria mercenaria*. *Mar. Ecol. Prog. Ser.* **1991**, *74*, 33–46. [CrossRef]

66. Sullivan, J.J. *Paralytic Shellfish Poisoning: Analytical and Biochemical Investigations*; University of Washington: Seattle, WA, USA, 1982.

67. Montojo, U.M.; Romero, M.L.; Borja, V.M.; Sato, S. Comparative PSP toxin accumulation in bivalves, *Paphia undulata* and *Perna viridis* in Sorsogon Bay, Philippines. In *Proceedings of the Seventh International Conference on Molluscan Shellfish Safety*; Lassus, P., Ed.; Ifremer: Nantes, France, 2010; pp. 49–55.

68. Contreras, A.M.; Marsden, I.D.; Munro, M.H.G. Physiological effects and biotransformation of PSP toxins in the New Zealand scallop, *Pecten novaezelandiae*. *J. Shellfish Res.* **2012**, *31*, 1151–1159. [CrossRef]

69. Bricelj, V.M.; Lee, J.M.; Cembella, A.D.; Anderson, D.M. Uptake kinetics of paralytic shellfish toxins from the dinoflagellate *Alexandrium fundyense* in the mussel *Mytilus edulis*. *Mar. Ecol. Prog. Ser.* **1990**, *63*, 177–188. [CrossRef]

70. Ichimi, K.; Suzuki, T.; Yamasaki, M. Non-selective retention of PSP toxins by the mussel *Mytilus galloprovincialis* feb with the toxic dinoflagellate *Alexandrium tamarense*. *Toxicon* **2001**, *39*, 1917–1921. [CrossRef]

71. Blanco, J.; Reyero, M.; Franco, J. Kinetics of accumulation and transformation of paralytic shellfish toxins in the blue mussel *Mytilus galloprovincialis*. *Toxicon* **2003**, *42*, 777–784. [CrossRef]

72. Medina-Elizalde, J.; García-Mendoza, E.; Turner, A.D.; Sánchez-Bravo, Y.A.; Murillo-Martínez, R. Transformation and depuration of paralytic shellfish toxins in the Geoduck clam *Panopea globosa* from the Northern Gulf of California. *Front. Mar. Sci.* **2018**, *5*, 335. [CrossRef]

73. Sullivan, J.J.; Iwaoka, W.T.; Liston, J. Enzymatic transformation of PSP toxins in the littleneck clam (*Protothaca staminea*). *Biochem. Biophys. Res. Commun.* **1983**, *114*, 465–472. [CrossRef]

74. Turner, A.D.; Lewis, A.M.; O'Neil, A.; Hatfield, R.G. Transformation of paralytic shellfish poisoning toxins in UK surf clams (*Spisula solida*) for targeted production of reference materials. *Toxicon* **2013**, *65*, 41–58. [CrossRef]

75. Haberkorn, H.; Lambert, C.; Le Goïc, N.; Moal, J.; Suquet, M.; Guéguen, M.; Sunila, I.; Soudant, P. Effects of *Alexandrium minutum* exposure on nutrition-related processes and reproductive output in oysters *Crassostrea gigas*. *Harmful Algae* **2010**, *9*, 427–439. [CrossRef]

76. Estrada, N.; de Jesús Romero, M.; Campa-Córdova, A.; Luna, A.; Ascencio, F. Effects of the toxic dinoflagellate, *Gymnodinium catenatum* on hydrolytic and antioxidant enzymes, in tissues of the giant lions-paw scallop *Nodipecten subnodosus*. *Comp. Biochem. Physiol. C Toxicol. Pharmacol.* **2007**, *146*, 502–510. [CrossRef]

77. Galimany, E.; Sunila, I.; Hégaret, H.; Ramón, M.; Wikfors, G.H. Pathology and immune response of the blue mussel (*Mytilus edulis* L.) after an exposure to the harmful dinoflagellate *Prorocentrum minimum*. *Harmful Algae* **2008**, *7*, 630–638. [CrossRef]

78. Hégaret, H.; da Silva, P.; Sunila, I.; Dixon, M.S.; Alix, J.; Shumway, S.E.; Wikfors, G.H.; Soudant, P. Perkinsosis in the Manila clam *Ruditapes philippinarum* affects responses to the harmful-alga, *Prorocentrum minimum*. *J. Exp. Mar. Biol. Ecol.* **2009**, *371*, 112–120. [CrossRef]

79. Galimany, E.; Sunila, I.; Hégaret, H.; Ramón, M.; Wikfors, G.H. Experimental exposure of the blue mussel (*Mytilus edulis*, L.) to the toxic dinoflagellate *Alexandrium fundyense*: Histopathology, immune responses, and recovery. *Harmful Algae* **2008**, *7*, 702–711. [CrossRef]
80. Abi-Khalil, C.; Lopez-Joven, C.; Abadie, E.; Savar, V.; Amzil, Z.; Laabir, M.; Rolland, J.L. Exposure to the paralytic shellfish toxin producer *Alexandrium catenella* increases the susceptibility of the oyster *Crassostrea gigas* to pathogenic vibrios. *Toxins* **2016**, *8*, 24. [CrossRef]
81. López, A.Z.; Cárdenas, L.; González, M.T. Metazoan symbionts of the yellow clam, *Mesodesma donacium* (Bivalvia), in Southern Chile: Geographical variations. *J. Parasitol.* **2014**, *100*, 797–804. [CrossRef]
82. MacQuarrie, S.P.; Bricelj, V.M. Behavioral and physiological responses to PSP toxins in *Mya arenaria* populations in relation to previous exposure to red tides. *Mar. Ecol. Prog. Ser.* **2008**, *366*, 59–74. [CrossRef]
83. Phillips, J.M.; Bricelj, V.M.; Mitch, M.; Cerrato, R.M.; MacQuarrie, S.; Connell, L.B. Biogeography of resistance to paralytic shellfish toxins in softshell clam, *Mya arenaria* (L.), populations along the Atlantic coast of North America. *Aquat. Toxicol.* **2018**, *202*, 196–206. [CrossRef] [PubMed]
84. Bricelj, V.M.; MacQuarrie, S.P.; Doane, J.A.E.; Connell, L.B. Evidence of selection for resistance to paralytic shellfish toxins during the early life history of soft-shell clam, *Mya arenaria*, populations. *Limmology Oceanogr.* **2010**, *55*, 2463–2475. [CrossRef]
85. Bricelj, V.M.; Connell, L.; Konoki, K.; Macquarrie, S.P.; Scheuer, T.; Catterall, W.A.; Trainer, V.L. Sodium channel mutation leading to saxitoxin resistance in Clams increases risk of PSP. *Nature* **2005**, *434*, 763–767. [CrossRef]
86. Anonymous. AOAC Official Method 959.08. Paralytic shellfish poison. Biological method. Final action. In *AOAC Official Methods for Analysis*; Truckses, M.W., Ed.; AOAC International: Gaithersburg, MD, USA, 2008; pp. 79–80.
87. van de Riet, J.; Gibbs, R.S.; Muggah, P.M.; Rourke, W.A.; MacNeil, J.D.; Quilliam, M.A. Liquid Chromatography Post-Column Oxidation (PCOX) method for the determination of Paralytic Shellfish Toxins in mussels, clams, and scallops: Collaborative study. *J. Aoac Int.* **2011**, *94*, 1154–1176. [PubMed]
88. Alexander, J.; Benford, D.; Boobis, A.; Ceccatelli, S.; Cravedi, J.P.; Domenico, A.D.; Doerge, D.; Dogliotti, E.; Edler, L.; Farmer, P.; et al. Scientific opinion of the panel on contaminants in the food chain on a request from the european commission of marine biotoxins in shellfish—saxitoxin group. *EFSA J.* **2010**, *8*, 1–76.
89. Shaw, B.L.; Battle, H.I. The gross microscopic anatomy of the digestive tract of the oyster *Crassostrea virginica* (Gmelin). *Can. J. Zool.* **1957**, *35*, 325–347. [CrossRef]
90. Howard, D.W.; Lewis, E.J.; Keller, B.J.; Smith, C.S. *Histological Techniques for Marine Bivalve Mollusks and Crustaceans*; NOAA Technical Memorandum NOS NCCOC: Oxford, UK, 2004.

toxins

MDPI

Article

Development and Application of Immunoaffinity Column Purification and Ultrahigh Performance Liquid Chromatography-Tandem Mass Spectrometry for Determination of Domoic Acid in Shellfish

Si Chen [1,2], Xiaojun Zhang [1,2,*], Zhongyong Yan [1,2], Yangyang Hu [3] and Yibo Lu [4]

[1]   Laboratory of aquatic product processing and quality safety, Marine Fisheries Research Institute of Zhejiang Province, Zhoushan 316100, Zhejiang, China; sichen_ns@zjou.edu.cn (S.C.); yanzhongyong@zjou.edu.cn (Z.Y.)
[2]   Marine and Fisheries Research Institute, Zhejiang Ocean University, Zhoushan 316000, Zhejiang, China
[3]   School of Food and Pharmacy, Zhejiang Ocean University, Zhoushan 316022, Zhejiang, China; yangyangHU6@hotmail.com
[4]   Jiangsu Meizheng Biotechnology Company Limited, Wuxi 214135, Jiangsu, China; luyibo_byl@163.com
*   Correspondence: zhangxj@zjou.edu.cn; Tel.: +86-580-2299896

Received: 12 December 2018; Accepted: 29 January 2019; Published: 1 February 2019

Abstract: Domoic acid (DA) is a neurotoxin associated with amnesic shellfish poisoning (ASP). Though LC coupled to tandem mass spectrometry (LC-MS/MS) has become the preferred method for DA determination, traditional sample pretreatment is still labor-intensive. In this study, a simple, efficient and selective method for LC-MS/MS analysis of DA in shellfish was established by optimizing clean-up procedures on a self-assembly immunoaffinity column (IAC). Shellfish was extracted with 75% methanol twice and diluted with phosphate buffered saline (PBS, 1:2). The mixture was purified on IAC as follows: preconditioned with PBS, loaded with sample, washed by 50% MeOH, and eluted with MeOH containing 2% ammonium hydroxide. Concentrated analyte was monitored by multiple reaction monitoring (MRM) using electrospray (ESI) positive ion mode throughout the LC gradient elution. Based on the post-extraction addition method, matrix effects for various shellfish matrices were found to be less than 8%. The developed method was fully validated by choosing mussel as the representative matrix. The method had a limit of detection (LOD) of 0.02 $\mu g \cdot g^{-1}$, showed excellent linear correlation in the range of 0.05–40 $\mu g \cdot g^{-1}$, and obtained ideal recoveries (91–94%), intra-day RSDs (6–8%) and inter-day RSDs (3–6%). The method was successfully applied to DA determination in 59 shellfish samples, with a detection rate of 10% and contaminated content of 0.1–14.9 $\mu g \cdot g^{-1}$.

Keywords: domoic acid; immunoaffinity column; purification; ultrahigh high performance liquid chromatography tandem mass spectrometry; shellfish

Key Contribution: A highly efficient and selective method was successfully developed and applied for DA determination in various shellfish matrices by coupling optimal IAC clean-up with Ultrahigh Performance Liquid Chromatography-Tandem Mass Spectrometry (UHPLC-MS/MS) analysis.

---

1. Introduction

Domoic acid (DA) is a glutamic acid analogue neurotoxin associated with amnesic shellfish poisoning (ASP) (Figure 1) [1]. The toxin was identified as the causative agent of the 1987 human poisoning incident in Canada, which occurred after consumption of contaminated blue mussels [2]. The human intoxication syndrome of ASP is characterized by gastrointestinal symptoms of vomiting, abdominal cramp, diarrhea, and neurological symptoms of headache, short-term memory loss, brain damage, and in the most severe cases, death [3]. DA is naturally produced by different species of *Pseudo-nitzschia* and other marine organisms such as red alga *Chondria armata*, then potentially accumulated in shellfish and other crustaceans through the food chain [4,5]. Reports of DA poisoning in wild animals, as well as DA contamination in coastal water have been published throughout the world [5–8]. To protect human health, Canada, European Union and United States have established that the upper limit of DA in the wet tissue of shellfish samples should not exceed 20 $\mu g \cdot g^{-1}$ [9].

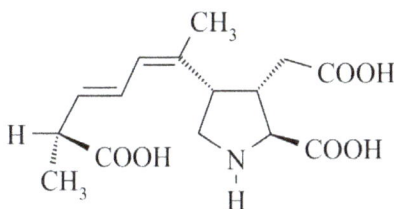

Figure 1. Chemical structure of domoic acid.

Currently, methods for DA identification and quantification include toxicity assay in mouse [10], enzyme-linked immunosorbent assay (ELISA) [11,12], thin layer chromatography (TLC) [13] and liquid chromatography (LC) [8,14–19]. The standard AOAC method, LC coupled with ultraviolet detection (LC-UV), has been widely employed and modified for DA monitoring, yet false positive results can occur when the samples contain interfering species [15]. Several LC coupled with fluorescence detector (LC-FLD) methods were proposed for DA determination in plankton and seawater through tedious derivatization procedures, but only two of these procedures have been successfully applied to shellfish matrices [16,17]. Due to its high sensitivity and selectivity, LC coupled to tandem mass spectrometry (LC-MS/MS) has become the preferred method for the confirmatory analysis of DA in shellfish. Possible matrix effects of shellfish make an efficient pretreatment procedure necessary before MS detection. Solid-phase extraction (SPE) based on strong anion exchange (SAX) has been employed as the most common clean-up method for biological samples [9,18,20]. Considering the fact that the practical operation of SAX could be labor-intensive, an immunoaffinity column (IAC) [21], molecularly imprinted solid-phase extraction (MISPE) [9], magnetic solid-phase extraction (MSPE) [22] and other SPE deformations are emerging as promising and innovative sample preparation techniques due to their specificity and convenience.

IAC clean-up utilizes the high specificity of imprinted antibodies to extract or concentrate target compounds from complex matrices [23]. With the strong advantages of reducing both interfering background and extraction time, this efficient and consistent technique has been widely applied as the alternative pretreatment method for routine SPE procedure to detect mycotoxins, veterinary drugs, pesticides and vitamins in food [24], yet only a few studies have been devoted to develop specific IAC to extract phycotoxins from marine organism tissue. Kawatsu et al were the first to report DA confirmation in Japanese mussels by LC coupled with IAC using an anti-DA monoclonal antibody as a ligand [21]. However, the proposed IAC seemed to be only compatible with phosphate-buffered saline (PBS) but organic solvents, and detailed information on method validation was lacked for possible further application. A tetrodotoxin-specific IAC was successfully developed and commercialized by our laboratory for LC-MS/MS determination of the potent neurotoxin in various marine organisms [25]. The aim of this work was to develop self-assembled IAC and corresponding procedures for DA purification from shellfish samples, which would be compatible with the extensively used methanolic extract and later analytical step of LC-MS/MS. The method was validated in terms of sensitivity, linearity, precision and repeatability by using mussel as the representative matrix, and then applied to DA determination in shellfish samples collected from local markets and culturing farms.

2. Results and Discussion

2.1. Optimization of UHPLC-MS/MS Conditions

Various DA isomers often co-occur at variable low levels in natural shellfish samples [26]. It is common practice to report the sum of DA and its major isomer C5′-*epi*-domoic acid (*epi*-DA) as the quantitative results of DA analysis [27]. Given the fact that DA exhibits higher toxicity than the isomers, and DA shares identical molecular weight and fragment ions with its isomers, selective extraction or complete resolution is necessary to avoid any potential overestimation during the LC-MS/MS analysis. In this study, IAC specific to the DA but not its isomers was developed and employed as the clean-up strategy, thereby simplifying subsequent instrumental analysis. For chromatographic separation, the mobile phase of acetonitrile/water and additive of formic acid/ammonium acetate were chosen. An LC gradient elution on an ACQUITY UPLC BEH C18 column with a high amount of organic component was utilized to obtain the best peak shape for the target analyte.

To obtain maximum abundance of molecular ions and generate higher sensitivity for analyte, optimization of MS/MS parameters was necessary. By direct infusing a DA standard solution (5.0 $\mu g \cdot mL^{-1}$) at a flow rate of 5 $\mu L \cdot min^{-1}$, a characteristic protonated adduct was obtained under full scan mode. The predominant peak corresponding to the $[M + H]^+$ ion at m/z 312.2 was chosen as the precursor ion. Collision energy was then introduced to generate product ion spectra of the selected $[M + H]^+$ ion using product ion scan mode for DA, and fragment ions at m/z 266.2 ($[M + H-HCOOH]^+$), m/z 248.2 ($[M + H-HCOOH-H_2O]^+$), m/z 161.2 ($[M + H-HCOOH-C_2H_3O_2N-H_2O-CH_2]^+$), m/z 220.1 ($[M + H-2HCOOH]^+$), m/z 193.1 ($[M + H-HCOOH-C_2H_3O_2N]^+$) were observed as the major and consistent ones. In order to optimize the MRM conditions for DA detection, product ion spectra were acquired at collision energies ranging from 8 eV to 40 eV at 4 eV intervals using optimized source conditions (Figure 2). Though fragment ion m/z 161.2 is less abundant but more specific, no significant difference was found in the MRM interference or background among transitions 312.2 > 161.2, 312.2 > 266.2 and 312.2 > 248.2. The most abundant and second most abundant fragment ions at m/z 266.2 and m/z 248.2 were selected for quantification purpose and qualitative confirmation, respectively.

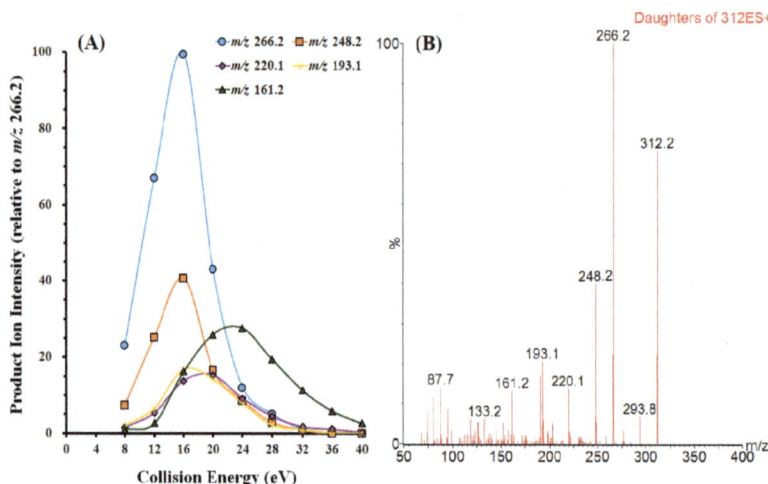

Figure 2. (**A**) The influence of collision energy on the product ion intensity and (**B**) Product ion MS2 spectra of $[M + H]^+$ ion (m/z 312.2) at a collision energy of 16 eV.

2.2. Optimization of Extracting Conditions

The most extensively used extraction procedures for DA involve 0.1 M HCl or aqueous methanol (MeOH) [28,29]. Previous studies suggested that low pH would induce the decomposition of DA [30,31], while extraction with 50% MeOH could provide a complete recovery of DA, as well as the other co-extracting yet interfering components from the sample matrix [29]. Higher percentages of organic solvent in the extraction system usually guarantee effective protein precipitation and lipid removal, which potentially enhance the column reusability by addressing the problem of sorbent clocking or blocking. To investigate the influences of the MeOH proportion and number of extractions on the DA recoveries, experiments were carried out by spiking DA-free shellfish samples at a level of 0.5 $\mu g \cdot g^{-1}$ before the extraction. The extracts were then purified with the IAC and analyzed by the UHPLC-MS/MS system. As shown in Table 1, single extraction with 50% or 75% MeOH was more effective than 90% MeOH for the DA recovery, exhibiting 11% and 17% higher efficacy, respectively. When the number of extractions increased from one to two, DA recovery for all the extractives exhibited an increase of 10–17%, and a similar changing trend was noticed among the 50%, 75% and 90% MeOH treated samples. Since the 50% methanolic extract of shellfish was relatively cloudier, which subsequently extended the sample loading process on IAC, DA extraction was performed twice with 75% MeOH in a subsequent study.

Table 1. Effect of the methanol percentage and number of extractions on the recoveries of DA (*n* = 3).

DA Recovery (%)	Number of Extraction	
	1	2
50% MeOH	82 ± 8	99 ± 8
75% MeOH	88 ± 3	102 ± 6
90% MeOH	71 ± 6	81 ± 7

2.3. Preparation of the IAC Column

The antibody specificity and column capacity are important factors that influence analyte recovery during the IAC-based purification process. In this study, DA-BSA (bovine serum albumin) conjugate was synthesized, and monoclonal antibody (Mab) against DA was successfully produced utilizing the hybridoma method [32]. Sepharose 4B was chosen as the ideal support matrix due to the following characteristics: it is water-insoluble but hydrophilic, can be easily activated, and has good physicochemical stability, high specificity and adsorption capacity. Both CNBr and NHS activated Sepharose 4B were employed for the antibody-matrix coupling to compare their efficiency. Different amounts of Mab were separately conjugated with these two matrices while the protein concentrations before and after the reactions were measured. As shown in Figure 3, when treated with the same amount of Mab (3–15 mg for 1 mL·gel), the coupling capacity of the CNBr-Sepharose 4B was 6–17% higher than that of the NHS-matrix. Conjugation of Sepharose 4B matrices with 6 mg Mab reached chemical saturation, among which the CNBr-activated one exhibited the highest coupling efficiency of 99 ± 2%. An excessive amount of antibody (>6 mg) might provide an increased blocking effect on the matrix surface, potentially reducing antigen-antibody binding and column capacity; therefore, 6 mg of Mab for per mL CNBr-Sepharose 4B was recommended for the chemical reaction in order to prevent ligand waste and maintain the column performance.

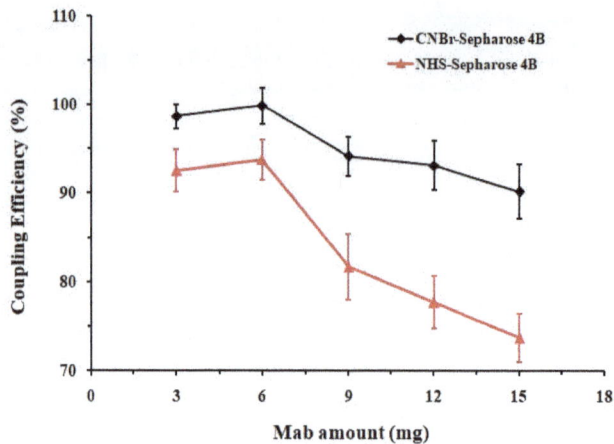

Figure 3. Coupling efficiency of CNBr- and NHS-activated Sepharose 4B (1 mL) conjugated with different amount of mAb (*n* = 3).

2.4. Characteristics of the IAC Column

Employing the high specificity of IAC as a sample pretreatment for LC-MS/MS offers a significant advantage over the existing methodology by providing the direct quantitative information of DA itself. CRM-ASP-Mus-d is a thermally sterilized homogenate of mussel tissue (*Mytilus edulis*) contaminated with DA and some of its isomers. Portions (1 mL) of methanolic extract were either directly evaporated by nitrogen or further cleaned up by IAC. These residues were then reconstituted by 1 mL of the initial mobile phase before the LC-MS/MS analysis (Figure 4). To separate DA from its isomers, a gradient cycle 7 min long was implemented with a 3-min isocratic segment (90% mobile phase B), a 0.5-min linear gradient to 10% B, a 1-min hold, a 0.5-min linear gradient to 90% B and a 2-min hold for column equilibration. As demonstrated by Figure 4, IAC assembled in this study permitted selective extraction of DA from its isomers and other interfering co-extracts in the shellfish sample, eliminating any possibility of overestimation or underestimation of the DA concentration. Studies have showed that DA can transform into *epi*-DA and other iso-DAs through heating, exposure to ultraviolet light or during long-term storage [33]. Since DA was solely extracted and detected, the IAC-based method therefore should preferably be used on fresh shellfish products as within these products, epimerization did not occur.

Column capacity and reusability depended on the concentration and activity of the immobilized Mab, as well as the chemical stability of the support material. The capacity was evaluated as mentioned in Section 4.4 and presented as micrograms of analyte relative to IAC bed volume ($\mu g \cdot mL^{-1}$). The column capacity of new IAC was found to be 6.4 $\mu g \cdot mL^{-1}$, which was sufficient for DA determination and suitable for shellfish products. Each time after DA purification, the IAC was washed and stored in PBS at 4 °C, and the column reusability was evaluated over 10 days at 1-day intervals. For the daily experiment, 1-mL portions of methanolic shellfish extract containing 50 ng DA were diluted with 2 mL PBS and then cleaned up with the optimized IAC procedure. As shown in Figure 5, though the recovery of IAC was consistent in the range of 89–97%, the column capacity decreased when the cycles of usage increased, especially during the third and fourth runs. Once the usage cycle exceeded a value of five, the IAC capacity dropped to 3.1 $\mu g \cdot mL^{-1}$. Considering the fact that the applicable linear curve was constructed with a concentration of 0.005–4 $\mu g \cdot mL^{-1}$ in this study, the IAC column therefore has the potential to be reused four times with column capacity above 4 $\mu g \cdot mL^{-1}$ to meet the method measure range.

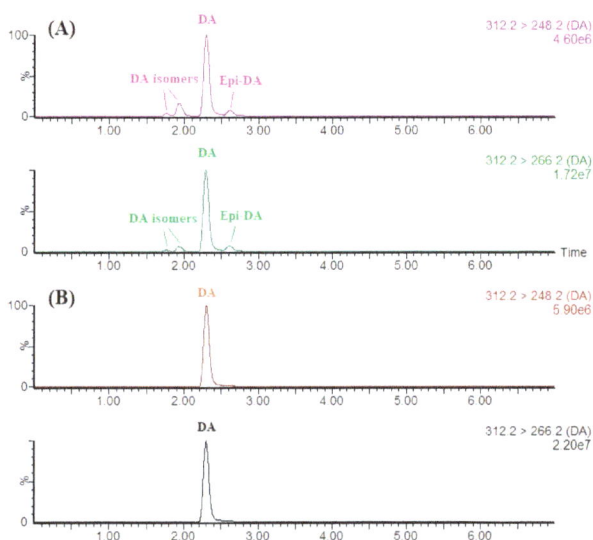

Figure 4. MRM analyses for DA and its isomers in 1-mL portions of CRM-ASP-Mus-d extract (**A**) before and (**B**) after IAC treatment (injection volume, 2 μL).

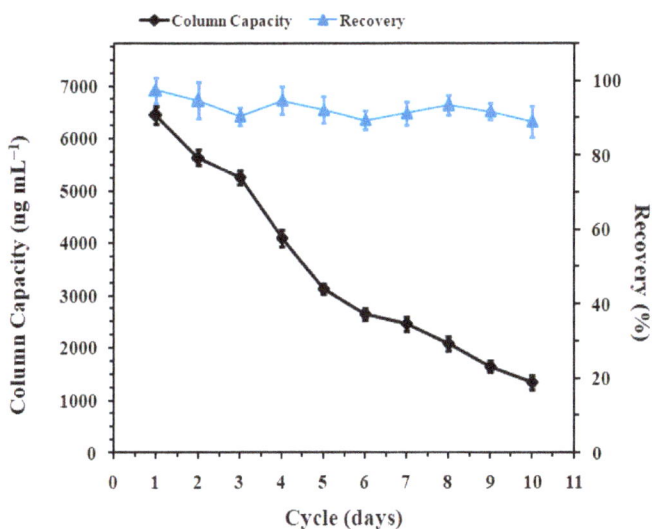

Figure 5. Evaluation on the capacity and recovery of the IAC column (*n* = 3).

2.5. Optimization of IAC Loading Conditions

Since organic solvent could denature the antibody and interfere with the antibody-antigen interaction, the primary task in this study was to reduce the MeOH concentration to an ideal level that allowed the analyte to be enriched by IAC without compromising the immunosorbent's performance. MeOH-PBS mixtures with different volumetric ratios at 1:9, 1:3 and 1:1 were tested as the loading solution. Specifically, 3 mL of each loading solution containing 50 ng DA was applied to the IAC at a constant flow rate of 2 mL·min^{-1}. The DA recoveries were found to be close to 95% when the MeOH percentage ranged from 10% to 25%, then showed an obvious decline as MeOH increased to 50% by volume. This result was in accordance with our previous work [25]. To achieve the best recovery and save pretreatment time, the MeOH amount of the sample extract was therefore diluted with PBS to 25% throughout the study.

2.6. Optimization of IAC Eluting Conditions

For the convenience of the later LC-MS/MS analysis, a small volume of organic solvent would be preferred to elute the analyte from the IAC. MeOH (10 mL) was initially tested by collecting each 1 mL eluate fraction. However, the results were not satisfactory, with DA recoveries of only 26% and 53% for the first and the second 5 mL eluate, respectively, which indicate that pure MeOH was not strong enough to break the antibody-antigen binding.

Considering the fact that further disruption of the binding interaction could be realized by raising or lowering the pH of the eluting solvent, and DA can decompose under low pH conditions [30,31], acidic (1% formic acid in MeOH, 5 mL) and alkali MeOH (1% and 2% ammonium hydroxide in MeOH, respectively, 5 mL) were selected as the eluant. For each eluting condition, the corresponding eluate fractions were collected and prepared by the same procedure mentioned above. As shown in Table 2, 2% ammonium hydroxide in MeOH gave the best recovery of 95%, and a volume of 3 mL was sufficient to elute the adsorbed DA completely from the IAC. Consequently, 3 mL of 2% ammonium hydroxide in MeOH was selected for the elution condition.

Table 2. Recoveries of DA under different eluting conditions ($n = 3$).

Elution Fraction	Recovery in Various Fractions (%)		
	1% Formic Acid in MeOH	1% Ammonium Hydroxide in MeOH	2% Ammonium Hydroxide in MeOH
fraction 1 (1 mL)	83	69	77
fraction 2 (1 mL)	4	15	16
fraction 3 (1 mL)	ND [b]	2	2
fraction 4 (1 mL)	ND	2	ND
fraction 5 (1 mL)	ND	ND	ND
total recovery [a]	87	88	95

[a] Total recovery is the sum of fraction 1 to fraction 5. [b] ND indicates not detected.

2.7. Optimization of IAC Washing Condition

Shellfish matrices are complex and differ in composition from one sample to another; a significant matrix effect and signal suppression were noticed during MS quantitation for DA in various shellfish extracts. Sample amount reduction, optimized sample preparation, and isotope internal standard or matrix-matched standards are usually used to compensate for these effects [34]. In this study, we aim to further remove the interfering substances from shellfish matrices by washing the IAC after the loading step to improve the accuracy of DA quantitation. Various aqueous MeOH percentages (10%, 20%, 30%, 40%, 50%, 60%, 70%, 80%, v/v) were selected and compared as the washing solvent. To investigate whether the MeOH concentration would interfere with the antibody-antigen bound, experiments were initially carried out by loading 3 mL of an MeOH-PBS (1:3, v/v) solution containing 50 ng DA into the IAC, followed by rinsing with 6 mL of the washing solvent mentioned above. The results showed that the DA recovery was still higher than 91% when the MeOH concentration in washing medium reached 50%, then underwent nearly 18% and over 80% decrease once the MeOH level exceeded 60% and 70%, respectively.

To test the efficiency and consistency of IAC in eliminating nonspecifically bound compounds from the shellfish matrices, additional experiments were carried out with methanolic extract of blank shellfish samples that were spiked with the same amount of DA standard. Portions (1 mL) of the spiked extract were diluted with PBS, passed through the IAC and then purified by a washing step with 6 mL of 10%, 30% and 50% MeOH. By using 10% or 30% MeOH as the washing solvent, DA recoveries for scallop (*Patinopecten yessoensis*), oyster (*Ostrea rivularis Gould*), mussel (*Mytilus edulis*), clam (*Scapharca subcrenata*) remained at 90 ± 7% or 90% ± 9%, 60 ± 3% or 65% ± 6%, 51 ± 8% or 52% ± 4%, 79 ± 2% or 71% ± 6%, respectively. For the above-mentioned samples washed by 50% MeOH, the DA recoveries were greatly improved, reaching 92 ± 7%, 91 ± 4%, 88 ± 5%, 93 ± 2%, respectively. Therefore, to maximize the elimination effect of interfering matrix components from the real shellfish extract, 6 mL of 50% MeOH was selected as the proper washing solvent.

2.8. Method Validation

The matrix effect for DA was evaluated based on the post-extraction addition method by comparing the peak area of a neat standard solution with that of a pretreated sample extract spiked with an equivalent amount of analyte. Experiments were carried out using scallop, oyster, mussel and clam spiked at two concentration levels (0.25 and 2.5 $\mu g \cdot g^{-1}$) as the representative matrices. Matrix effects below 8% were obtained for all the shellfish samples, indicating no significant ion suppression or enhancement observed using the developed method.

Considering the fact that mussel is the main economic shellfish in the local marine aquaculture industry and mussel homogenate exhibited most serious matrix effect for MS detection during previous IAC optimization, the method was further validated by using mussel as the representative matrix. Validation parameters included the linear range, correlation coefficient, limit of detection (LOD), limit of quantitation (LOQ), intra-day and inter-day precisions and repeatability. A calibration curve was constructed with a concentration of 0.005–4 $\mu g \cdot mL^{-1}$ DA standard solution (namely, 0.05–40 $\mu g \cdot g^{-1}$ for mussel sample), and the linearity was good with correlation coefficients (R^2) higher than 0.998.

LC-UV is the usual instrumental method that allows detection of DA at the concentration level of the current regulatory limit with the need for further confirmatory analysis [15]. Studies involving trace analysis of DA in shellfish and early detection of DA toxicity events usually imply additional analytical methods with higher sensitivity and qualitative function. From the method performance of higher-level spiked samples, it was possible to accurately determine the LOD and LOQ for DA in mussel tissue. Repeated analysis of 0.05 $\mu g \cdot g^{-1}$ and lower spiked samples were carried out to see if the S/N value of quantitative transition m/z 312.2 > 266.2 could be measured. The LODs (S/N = 3) and LOQs (S/N = 10) for each shellfish matrix were then found to be comparable, reaching respective values of 0.02 and 0.05 $\mu g \cdot g^{-1}$ with a 2 μL injection volume.

Recoveries were performed by the standard addition method. Experiments were carried out over 5 days with a DA-free mussel homogenate spiked at three different levels (0.25, 2.5 and 25 $\mu g \cdot g^{-1}$) to assess accuracy, intra-assay and inter-assay precision (Table 3). DA recoveries for spiked samples ranged from 91–94% with intra-day RSDs of 6–8% and inter-day RSDs of 3–6%. Using the certified reference material CRM-ASP-Mus-d (49 ± 3 $\mu g \cdot g^{-1}$), recoveries of DA were 91 ± 7% (n = 3). Typical LC-MS/MS chromatograms of the blank sample, standard solution, spiked sample, and naturally DA-contaminated sample are shown in Figure 6.

Table 3. Accuracy, intra-assay and inter-assay precision of the developed method using a blank mussel spiked at three different levels.

Spiked Level ($\mu g \cdot g^{-1}$)	Intra-Day RSD (%, n = 5)	Inter-Day RSD (%, n = 3)	Recovery ± SD (%, n = 5)
0.25	7	6	91 ± 3
2.5	6	3	94 ± 5
25	8	5	92 ± 6

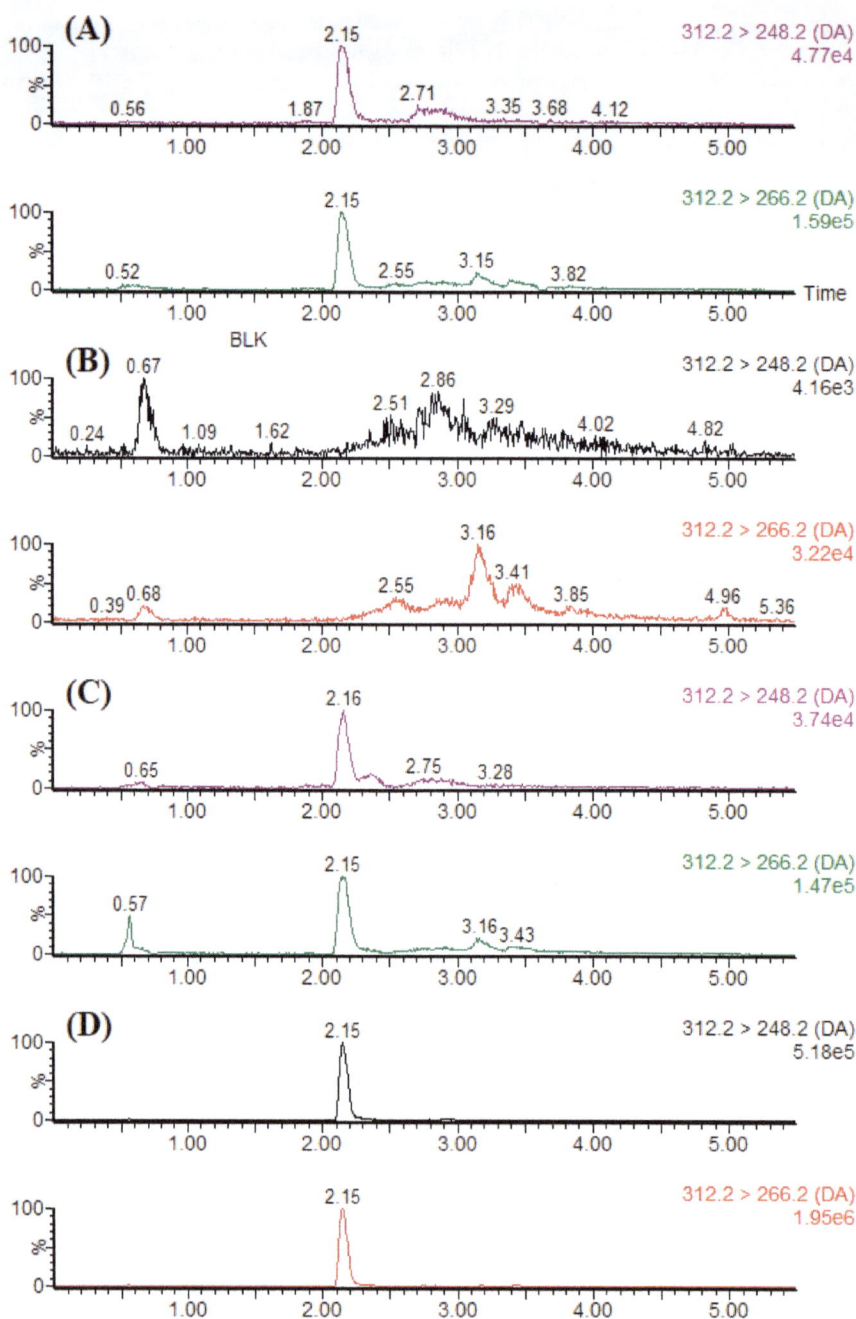

Figure 6. Representative MRM chromatogram of DA obtained from (**A**) 25.0 ng·mL^{-1} of standard solution; (**B**) blank mussel sample; (**C**) mussel sample spiked at a concentration of 0.25 µg·g^{-1}; (**D**) naturally DA-contaminated shellfish sample.

2.9. Application to Real Samples

For live and raw bivalve molluscs, Codex Stan 292-2008 stipulates that the minimum applicable range, LOD and LOQ for DA determination method should meet the respective criteria of 14–26 $\mu g \cdot g^{-1}$, 2 $\mu g \cdot g^{-1}$ and 4 $\mu g \cdot g^{-1}$ [35]. When the traditional method employs SAX as the clean-up strategy, matrix-matched standards should usually be constructed for all the tested samples for accurate quantitation [20,36]. Serving as simpler and more efficient clean-up procedures, SPE deformations such as IAC, MISPE and MSPE have gained an increasing amount of attention for DA analysis in shellfish samples. In Table 4, a comprehensive comparison of the proposed method with other existing methods for DA determination was summarized in term of extraction, clean-up, LOD, linear range and recovery. Zhang et al [22] proposed an MSPE method for LC-MS/MS analysis of DA in shellfish samples by synthesizing $Fe_3O_4 \cdot SiO_2 \cdot UiO$-66 magnetic microspheres, and Regueiro et al [37] utilized online coupling of weak anion exchange SPE and LC-UV-MS/MS to realize sensitive determination of DA in shellfish. Though LODs of the analyte in these two studies were reported at the level of $pg \cdot g^{-1}$ and sub $ng \cdot g^{-1}$, respectively, linear ranges constructed in the corresponding methods both fell outside of the one regulated by Codex Stan 292-2008. Not only were the recovery of IAC and detection limit for DA in this work comparable to other relevant reports involving deformed SPE procedures, but the performances of the developed method fulfilled the stipulation in Codex Stan 292-2008.

A total of 59 shellfish samples collected from five eastern coastal provinces between April and November 2017 were analyzed for the presence of DA. The detected species included mussel (*Mytilus edulis*), ark shell (*Scapharca subcrenata*), blood clam (*Tegillarca granosa*), different types of scallops (*Patinopecten yessoensis, Chlamys farreri, Argopecten irradians*) and oyster (*Ostrea rivularis Gould, Ostrea gigas Thunberg*) from local markets and culturing farms. DA has been detected in a number of different marine bivalves in the East and South China Seas, with inconsistent detection rate between 0–82% and a contaminated content of 0.02–18.2 $\mu g \cdot g^{-1}$ [7,30]. In this survey, the detection rate for DA reached 10%. Among the analyzed samples (Table 5), the largest percentage of toxic samples was from zhikong scallop (*Chlamys farreri*). Two positive zhikong scallops from the local market contained the highest DA concentrations of 14.9 and 11.8 $\mu g \cdot g^{-1}$, while the locally cultured one was found to have a low value of 0.7 $\mu g \cdot g^{-1}$. Other contaminated samples included two bay scallops (*Argopecten irradians*) from Liaoning province and one jinjiang oyster (*Ostrea rivularis Gould*) from Guangdong province, ranging from 0.1–9.3 $\mu g \cdot g^{-1}$. In all the analyzed shellfish, the DA appeared to be present at a level below the regulatory limit (20 $\mu g \cdot g^{-1}$ in edible tissue), but the results obtained in this study show a warning that the toxin still needs to be continuously monitored in the investigated bivalves, especially the ones from northeast costal area of China.

Table 4. A comparison between previously reported methods and the proposed method for LC determination of DA in shellfish.

Instrumental Method	Mode	Extraction	Clean-Up	LOD	Linear Range	Recovery	Reference
LC-UV	Offline	HCl	immunoaffinity column	-d	-	85–90%	[21]
LC-MS/MS	Offline	Aqueous MeOH	Strong anion SPE	$0.014\ \mu g \cdot mL^{-1}$	0.025–$10\ \mu g \cdot mL^{-1}$	92%	[18]
LC-UV	Offline	Aqueous MeOH	MIP [a] SPE	$0.1\ \mu g \cdot mL^{-1}$	0.5–$25\ \mu g \cdot mL^{-1}$	93–97%	[9]
LC-UV	Online	Aqueous MeOH	MIP monolithic column	$0.076\ \mu g \cdot mL^{-1}$	-	89–91%	[38]
LC-MS/MS	Offline	Aqueous MeOH	$Fe_3O_4 \cdot SiO_2 \cdot$ UiO-66 MSPE [b]	$1.45\ pg \cdot mL^{-1}$	2–$1000\ pg \cdot mL^{-1}$	93–107%	[22]
LC-UV-MS/MS	Online	Aqueous MeOH	weak anion SPE	$0.3\ ng \cdot g^{-1}$ (MS/MS) 4–$10\ ng \cdot g^{-1}$ (UV)	0.05–$100\ ng \cdot mL^{-1}$ (MS/MS) 0.25–$200\ ng \cdot mL^{-1}$ (UV)	94–102%	[37]
LC-MS/MS	Offline	methanol/acetone	Florisil (Extraction with PLE [c])	$0.2\ \mu g \cdot g^{-1}$	0.05–$5\ \mu g \cdot mL^{-1}$	81–95%	[20]
LC-MS/MS	Offline	Aqueous MeOH	immunoaffinity column	$0.02\ \mu g \cdot g^{-1}$	0.05–$40\ \mu g \cdot g^{-1}$	91–94%	This work

[a] MIP indicates molecularly imprinted polymer. [b] MSPE indicates magnetic solid-phase extraction. [c] PLE indicates pressurised liquid extraction. [d] - indicates not mentioned.

Table 5. DA Concentrations ($\mu g \cdot g^{-1}$) in 59 shellfish samples collected from eastern coastal provinces in China between April and November 2017.

Species	Total Number of Samples Analyzed	No. Positive Samples	Sampling Location	Sampling Time	DA Level ($\mu g \cdot g^{-1}$)
yesso scallop (*Patinopecten yessoensis*)	4	0			ND [a]
zhikong scallop (*Chlamys farreri*)	12	3			
		No. 1	Huludiao, Liaoning/local market	September 2017	11.8
		No. 2	Huludiao, Liaoning/local market	September 2017	14.9
		No. 3	Xingcheng, Liaoning/culturing farm	September 2017	0.7
bay scallop (*Argopecten irradians*)	10	2			
		No. 1	Huludiao, Liaoning/local market	September 2017	9.3
		No. 2	Xingcheng, Liaoning/culturing farm	September 2017	3.2
jinjiang oyster (*Ostrea rivularis Gould*)	9	1			
		No. 1	Shantou, Guangdong/culturing farm	October 2017	0.1
long oyster (*Ostrea gigas Thunberg*)	6	0			ND
mussel (*Mytilus edulis*)	8	0			ND
ark shell (*Scapharca subcrenata*)	5	0			ND
blood clam (*Tegillarca granosa*)	5	0			ND

[a] ND indicates not detected.

3. Conclusions

In this study, a highly efficient and selective method for the determination of DA in shellfish samples was developed based on IAC purification prior to UHPLC-MS/MS analysis. Clean-up procedures including loading, eluting and washing conditions on a self-assembled IAC were systematically optimized, which allowed accurate and sensitive DA quantitation without a significant matrix effect. The LOD, LOQ and applicable range of this method were 0.02 µg·g^{-1}, 0.05 µg·g^{-1} and 0.05–40 µg·g^{-1}, respectively. DA recoveries ranged from 91–94% with intra-day RSDs of 6–8% and inter-day RSDs of 3–6%. Method performances in this work fulfilled the regulation in Codex Stan 292-2008 and were comparable to other relevant reports. In a limited DA survey, various shellfish samples were successfully analyzed, and zhikong scallop (*Chlamys farreri*) was found to be the most toxic species. Results for method validation and application indicated that this IAC has a promising prospect for further development and commercialization as a pretreatment kit.

4. Materials and Methods

4.1. Chemicals and Reagents

A high-purity standard solution (purity higher than 98%) containing 103.3 ± 2.5 µg·mL^{-1} DA and certified reference material containing 49 ± 3 µg·g^{-1} DA (CRM-ASP-Mus-d) were purchased from the Canadian National Research Council (Halifax, Canada). Methanol (MeOH), acetonitrile (ACN), ammonium acetate and formic acid of LC grade were purchased from Sigma-Aldrich (St. Louis, MO, USA). Ultrapure water used in the preparation of reagent solutions and the mobile phase was obtained from a Milli-Q water purification system (Millipore, Bedford, MA, USA). All the other chemicals and solvents were of analytical reagent grade and were obtained from Shanghai Chemical Reagent Co. (Shanghai, China).

4.2. Standard Solutions and Buffers

DA stock standard solution (25 µg·mL^{-1}) was prepared by dilution in MeOH and stored at 4 °C. The working standard solutions were prepared by 2 or 5 volumetric serial dilutions of the stock solution using ultrapure water containing 0.1% (*v*/*v*) formic acid and 2 mM ammonium acetate/acetonitrile (9:1, *v*/*v*) as the diluent. Phosphate buffered saline (PBS, pH 7.4) was prepared by dissolving 1.09 g of KH$_2$PO$_4$·2 H$_2$O, 6.45 g of Na$_2$HPO$_4$·12 H$_2$O and 4.25 g of NaCl in 500 mL of ultrapure water.

4.3. Preparation of Immunoaffinity Column

Cynogen bromide (CNBr) or N-hydroxy-succinimide (NHS) activated Sepharose 4B powder (1 g) was swelled within hydrochloric acid (HCl, 1 M, 200 mL) and poured into a sintered glass funnel (40–60 mm). The gel was washed with the coupling buffer (0.1 M NaHCO$_3$, 0.5 M NaCl, pH 8.3, 1 L), and a 3 mL aliquot was coupled with DA monoclonal antibody (10 mg Mab mL^{-1} dissolved in the coupling buffer, 3 mL) in a stoppered flask and incubated on a shaker (120 rpm) at room temperature for 2 h. Subsequently, the mixture was washed with the coupling buffer (60 mL) to remove the free Mabs. The eluant was collected to determine the antibody amount by the Bradford protein assay method and to calculate the coupling efficiency. The unreacted active groups on the sorbents were capped with the blocking buffer (0.1 M Tris-HCl, pH 8.0) at room temperature for 2 h, then washed with 0.1 M acetate buffer (containing 0.5 M NaCl, pH 4.0) and 0.1 M Tris-HCl buffer (containing 0.5 M NaCl, pH 8.0). Finally, the Mab-coupled gel was equilibrated with 3 mL of PBS (0.01 M, pH 7.4), and then a 1 mL aliquot was transferred to a 3 mL SPE column and stored in PBS containing 0.01% (*w*/*v*) sodium azide at 4 °C.

4.4. Determination of Column Capacity

A relatively large amount (8 μg) of DA was spiked in PBS (3 mL) containing 25% MeOH. The solutions were then loaded into the IAC column (preconditioned with 3 mL of PBS) at a constant flow rate of 2 mL·min^{-1}. The saturated column was washed with 6 mL of aqueous 50% MeOH and then eluted with 3 mL of MeOH containing 2% ammonium hydroxide. The eluate was evaporated to dryness under a stream of nitrogen at 40 °C in a water bath. The residue was reconstituted by 1 mL of ultrapure water containing 0.1% (v/v) formic acid and 2 mM ammonium acetate/acetonitrile (9:1, v/v, 1 mL) and filtered through a 0.22 μm PTFE filter into an autosampler vial for later analysis.

4.5. Sample Preparation

Homogenized shellfish tissue (2.00 ± 0.01 g) was weighed into a 50 mL polypropylene centrifuge tube. Then, 8 mL of aqueous 75% MeOH was added. The mixture was vortexed for 90 s, ultrasonicated for 10 min, and centrifugated at 6000× g for 5 min. The resulting supernatant was transferred into a new tube and the pellets re-extracted as described above. The volume of the combined supernatants was adjusted to 20 mL with the extracting solvent, and 1 mL of the sample extract was diluted with PBS at a ratio of 1:2 to adjust the final MEOH concentration to 25%.

The IAC column was preconditioned with PBS (3 mL) prior to sample loading. The analyte was loaded onto the IAC, washed with 6 mL of aqueous 50% MeOH, and then eluted with 3 mL of MeOH containing 2% ammonium hydroxide at a constant flow rate of 2 mL·min^{-1}. The eluate was evaporated to dryness under a stream of nitrogen at 40 °C in a water bath. The residue was reconstituted by 1 mL of ultrapure water containing 0.1% (v/v) formic acid and 2 mM ammonium acetate/acetonitrile (9:1, v/v, 1 mL) and filtered through a 0.22 μm PTFE filter into an autosampler vial for later analysis. The columns, after elution, were immediately regenerated by PBS (50 mL) and stored in PBS containing 0.01% (w/v) sodium azide at 4 °C for subsequent use.

4.6. UHPLC-MS/MS Conditions

The DA analysis was carried out on an ACQUITY UPLC system coupled with a Xevo TQ-S triple-quadrupole mass spectrometer (Waters, Milford, MA, USA). The LC separation was performed on an ACQUITY UPLC BEH C18 column (50 mm × 2.1 mm I.D., 1.7 μm particle size) at 40 °C. The injection volume was 2 μL. Acetonitrile (A) and ultrapure water containing 0.1% (v/v) formic acid and 2 mM ammonium acetate (B) were used as the mobile phases, and the flow rate was 0.2 mL·min^{-1} throughout the analysis. The chromatographic resolution for DA and its isomers was set as follows (t is time and subscript numbers are the time in minute): t_0, B = 90; t_3, B = 90; $t_{3.5}$, B = 10; $t_{4.5}$, B = 10; t_5, B = 90; t_7, B = 90. The gradient for the sample cleaned up by IAC was set as follows: t_0, B = 90; t_1, B = 90; t_3, B = 10; t_4, B = 10; $t_{4.1}$, B = 90; $t_{5.5}$, B = 90.

Analyte was detected by multiple reaction monitoring (MRM) using electrospray (ESI) positive ion mode. The MS/MS detection parameters were optimized by standard infusion and set as follows: capillary voltage, 3.0 kV; source temperature, 150 °C; desolvation temperature, 600 °C; cone gas flow, 150 L·h^{-1}; desolvation gas flow, 1000 L·h^{-1}; cone voltage, 12 V; collision energy, 16 eV. The precursor/product ion m/z 312.2 > 266.2 and m/z 312.2 > 248.2 were used for quantitative determination and qualitative confirmatory, respectively.

4.7. Data Analysis

Data were acquired and processed using MassLynx 4.1 and QuanLynx software (Waters, Milford, MA, USA). The DA concentrations in samples (μg·g^{-1}) were calculated directly from the area responses using a linear seven-point calibration ranged from 0.005–4 μg·mL^{-1}. The confirmatory guidelines for DA were in agreement with the performance criteria of European Commission (EC) Decision 2002/657/EC [39]. The maximum permitted tolerances for the relative retention time and relative ion intensities (ion ratio) should be within ± 2.5% and ± 25%, respectively.

Author Contributions: S.C. designed and led the research and wrote the paper. S.C. and X.Z. analyzed the data. X.Z. reviewed the paper. S.C. and Z.Y. developed and performed the methodology. Z.Y. and Y.H. validated the method and the results. Y.H. and Y.L. performed the sample analysis. Y.L. contributed the materials and reagents. All authors read and approved the final manuscript.

Funding: This research was funded by the Technical Support Team Projects in Zhejiang Province for Quality and Safety of Aquatic Products, grant number QS2017005 and the Science and Technology Program of Zhoushan City, grant number 2017C32073.

Acknowledgments: The authors would like to thank their colleagues for their valuable technical assistance.

Conflicts of Interest: The authors declare no conflict of interest.

References

1. Ramsdell, J.S. The molecular and integrative basis to domoic acid toxicity. In *Phycotoxins: Chemistry and Biochemistry*, 1st ed.; Botana, L.M., Ed.; Blackwell Publishing: Ames, IA, USA, 2007; Volume 13, pp. 223–250. [CrossRef]

2. Bates, S.S.; Bird, C.J.; Freitas, A.S.W.D.; Foxall, R.; Gilgan, M.; Hanic, L.A.; Johnson, G.R.; McCulloch, A.W.; Odense, P.; Pocklington, R.; et al. Pennate Diatom Nitzschia pungens as the Primary Source of Domoic Acid, a Toxin in Shellfish from Eastern Prince Edward Island, Canada. *Can. J. Fish. Aquat. Sci.* **1989**, *46*, 1203–1215. [CrossRef]

3. Perl, T.M.; Bedard, L.; Kosatsky, T.; Hockin, J.C.; Todd, E.C.; Remis, R.S. An outbreak of toxic encephalopathy caused by eating mussels contaminated with domoic acid. *N. Engl. J. Med.* **1990**, *322*, 1775–1780. [CrossRef] [PubMed]

4. Barbaro, E.; Zangrando, R.; Rossi, S.; Cairns, W.R.L.; Piazza, R.; Corami, F.; Barbante, C.; Gambaro, A. Domoic acid at trace levels in lagoon waters: Assessment of a method using internal standard quantification. *Anal. Bioanal. Chem.* **2013**, *405*, 9113–9123. [CrossRef] [PubMed]

5. Lefebvre, K.A.; Robertson, A. Domoic acid and human exposure risks: A review. *Toxicon* **2010**, *56*, 218–230. [CrossRef] [PubMed]

6. Trainer, V.L.; Hickey, B.M.; Bates, S.S. Toxic diatoms. In *Oceans and Human Health: Risks and Remedies from the Sea*, 1st ed.; Walsh, P.J., Smith, S.L., Fleming, L.E., Solo-Gabriele, H.M., Gerwick, W.H., Eds.; Academic Press: Burlington, VT, USA, 2008; Volume 14, pp. 219–237.

7. Li, Y.; Huang, C.X.; Xu, G.S.; Lundholm, N.; Teng, S.T.; Wu, H.; Tan, Z. Pseudo-nitzschia simulans sp. nov. (Bacillariophyceae), the first domoic acid producer from Chinese waters. *Harmful Algae* **2017**, *67*, 119–130. [CrossRef] [PubMed]

8. Duxbury, M. Liquid chromatographic determination of amnesic shellfish poison in mussels. *J. Chem. Educ.* **2000**, *77*, 1319. [CrossRef]

9. Zhou, W.H.; Guo, X.C.; Zhao, H.Q.; Wu, S.X.; Yang, H.H.; Wang, X.R. Molecularly imprinted polymer for selective extraction of domoic acid from seafood coupled with high-performance liquid chromatographic determination. *Talanta* **2011**, *84*, 777–782. [CrossRef] [PubMed]

10. Campbell, K.; McNamee, S.E.; Huet, A.C.; Delahaut, P.; Vilarino, N.; Botana, L.M.; Poli, M.; Elliott, C.T. Evolving to the optoelectronic mouse for phycotoxin analysis in shellfish. *Anal. Bioanal. Chem.* **2014**, *406*, 6867–6881. [CrossRef] [PubMed]

11. Tsao, Z.J.; Liao, Y.C.; Liu, B.H.; Su, C.C.; Yu, F.Y. Development of a monoclonal antibody against domoic acid and its application in enzyme-linked immunosorbent assay and colloidal gold immunostrip. *J. Agric. Food. Chem.* **2007**, *55*, 4921–4927. [CrossRef] [PubMed]

12. Hesp, B.R.; Harrison, J.C.; Selwood, A.I.; Holland, P.T.; Kerr, D.S. Detection of domoic acid in rat serum and brain by direct competitive enzyme-linked immunosorbent assay (cELISA). *Anal. Bioanal. Chem.* **2005**, *383*, 783–786. [CrossRef] [PubMed]

13. Quilliam, M.A.; Thomas, K.; Wright, J.L. Analysis of domoic acid in shellfish by thin-layer chromatography. *Nat. Toxins* **1998**, *6*, 147–152. [CrossRef]

14. Vilarino, N.; Louzao, M.C.; Fraga, M.; Rodriguez, L.P.; Botana, L.M. Innovative detection methods for aquatic algal toxins and their presence in the food chain. *Anal. Bioanal. Chem.* **2013**, *405*, 7719–7732. [CrossRef] [PubMed]

15. Hess, P.; McGovern, E.; McMahon, T.; Morris, S.; Stobo, L.A.; Brown, N.A.; Gallacher, S.; McEvoy, J.D.G.; Kennedy, G.; Young, P.B.; et al. LC-UV and LC-MS methods for the determination of domoic acid. *Trends Analyt. Chem.* **2005**, *24*, 358–367. [CrossRef]

16. James, K.J.; Gillman, M.; Lehane, M.; Gago-Martinez, A. New fluorimetric method of liquid chromatography for the determination of the neurotoxin domoic acid in seafood and marine phytoplankton. *J. Chromatogr. A* **2000**, *871*, 1–6. [CrossRef]

17. Maroulis, M.; Monemvasios, I.; Vardaka, E.; Rigas, P. Determination of domoic acid in mussels by HPLC with post-column derivatization using 4-chloro-7-nitrobenzo-2-oxa-1,3-diazole (NBD-Cl) and fluorescence detection. *J. Chromatogr. B Analyt. Technol. Biomed. Life Sci.* **2008**, *876*, 245–251. [CrossRef] [PubMed]

18. Furey, A.; Lehane, M.; Gillman, M.; Fernandez-Puente, P.; James, K.J. Determination of domoic acid in shellfish by liquid chromatography with electrospray ionization and multiple tandem mass spectrometry. *J. Chromatogr. A* **2001**, *938*, 167–174. [CrossRef]

19. Tor, E.R.; Puschner, B.; Whitehead, W.E. Rapid determination of domoic acid in serum and urine by liquid chromatography-electrospray tandem mass spectrometry. *J. Agric. Food Chem.* **2003**, *51*, 1791–1796. [CrossRef] [PubMed]

20. Pardo, O.; Yusa, V.; Leon, N.; Pastor, A. Development of a pressurised liquid extraction and liquid chromatography with electrospray ionization-tandem mass spectrometry method for the determination of domoic acid in shellfish. *J. Chromatogr. A* **2007**, *1154*, 287–294. [CrossRef] [PubMed]

21. Kawatsu, K.; Hamano, Y.; Noguchi, T. Determination of domoic acid in Japanese mussels by enzyme immunoassay. *J. AOAC Int.* **2000**, *83*, 1384–1386.

22. Zhang, W.; Yan, Z.; Gao, J.; Tong, P.; Liu, W.; Zhang, L. Metal-organic framework UiO-66 modified magnetite@silica core-shell magnetic microspheres for magnetic solid-phase extraction of domoic acid from shellfish samples. *J. Chromatogr. A* **2015**, *1400*, 10–18. [CrossRef]

23. Weller, M.G. Immunochromatographic techniques—A critical review. *Fresenius J. Anal. Chem.* **2000**, *366*, 635–645. [CrossRef] [PubMed]

24. Senyuva, H.Z.; Gilbert, J. Immunoaffinity column clean-up techniques in food analysis: A review. *J. Chromatogr. B Analyt. Technol. Biomed. Life Sci.* **2010**, *878*, 115–132. [CrossRef] [PubMed]

25. Zhang, X.; Yan, Z.; Wang, Y.; Jiang, T.; Wang, J.; Sun, X.; Guo, Y. Immunoaffinity chromatography purification and ultrahigh performance liquid chromatography tandem mass spectrometry determination of tetrodotoxin in marine organisms. *J. Agric. Food Chem.* **2015**, *63*, 3129–3134. [CrossRef] [PubMed]

26. Wright, J.L.C.; Falk, M.; McInnes, A.G.; Walter, J.A. Identification of isodomoic acid D and two new geometrical isomers of domoic acid in toxic mussels. *Can. J. Chem.* **1990**, *68*, 22–25. [CrossRef]

27. Beach, D.G.; Walsh, C.M.; McCarron, P. High-throughput quantitative analysis of domoic acid directly from mussel tissue using Laser Ablation Electrospray Ionization-tandem mass spectrometry. *Toxicon* **2014**, *92*, 75–80. [CrossRef] [PubMed]

28. Lawrence, J.F.; Charbonneau, C.F.; Ménard, C. Liquid chromatographic determination of domoic acid in mussels, using AOAC paralytic shellfish poison extraction procedure: Collaborative study. *J. Assoc. Off. Anal. Chem.* **1991**, *74*, 68–72. [PubMed]

29. Quilliam, M.A.; Xie, M.; Hardstaff, W.R. Rapid extraction and cleanup for liquid chromatographic determination of domoic acid in unsalted seafood. *J. AOAC Int.* **1995**, *78*, 543–554.

30. Jeffery, B.; Barlow, T.; Moizer, K.; Paul, S.; Boyle, C. Amnesic shellfish poison. *Food Chem. Toxicol.* **2004**, *42*, 545–557. [CrossRef]

31. Vale, P.; Sampayo, M.A. Evaluation of extraction methods for analysis of domoic acid in naturally contaminated shellfish from Portugal. *Harmful Algae* **2002**, *1*, 127–135. [CrossRef]

32. Erlanger, B.F.; Borek, F.; Beiser, S.M.; Lieberman, S. Steroid-protein conjugates. II. Preparation and characterization of conjugates of bovine serum albumin with progesterone, deoxycorticosterone, and estrone. *J. Biol. Chem.* **1959**, *234*, 1090–1094.

33. Farabegoli, F.; Blanco, L.; Rodríguez, L.P.; Vieites, J.M.; Cabado, A.G. Phycotoxins in marine shellfish: Origin, occurrence and effects on humans. *Mar. Drugs* **2018**, *16*, 188. [CrossRef] [PubMed]

34. Chambers, E.; Wagrowski-Diehl, D.M.; Lu, Z.; Mazzeo, J.R. Systematic and comprehensive strategy for reducing matrix effects in LC/MS/MS analyses. *J. Chromatogr. B Analyt. Technol. Biomed. Life Sci.* **2007**, *852*, 22–34. [CrossRef] [PubMed]

35. Food and Agriculture Organization of the United Nations/World Health Organization. *CODEX STAN 292-2008. Codex Standard for Live and Raw Bivalve Molluscs*; Codex Alimentarius Commission: Rome, Italy, 2008.
36. He, Y.; Fekete, A.; Chen, G.; Harir, M.; Zhang, L.; Tong, P.; Schmitt-Kopplin, P. Analytical approaches for an important shellfish poisoning agent: Domoic Acid. *J. Agric. Food Chem.* **2010**, *58*, 11525–11533. [CrossRef] [PubMed]
37. Regueiro, J.; Martín-Morales, E.; Álvarez, G.; Blanco, J. Sensitive determination of domoic acid in shellfish by on-line coupling of weak anion exchange solid-phase extraction and liquid chromatography–diode array detection–tandem mass spectrometry. *Food Chem.* **2011**, *129*, 672–678. [CrossRef] [PubMed]
38. Yang, F.; Wang, R.; Na, G.; Yan, Q.; Lin, Z.; Zhang, Z. Preparation and application of a molecularly imprinted monolith for specific recognition of domoic acid. *Anal. Bioanal. Chem.* **2018**, *410*, 1845–1854. [CrossRef] [PubMed]
39. European Commission. Commission Decision 2002/657/EC implementing Council Directive 96/23/EC concerning the performance of analytical methods and the interpretation of results. *Off. J. Eur. Communities* **2002**, *L221*, 8–36.

toxins

MDPI

Article

Combined Cytotoxicity of the Phycotoxin Okadaic Acid and Mycotoxins on Intestinal and Neuroblastoma Human Cell Models

Aiko Hayashi [1,*], Juan José Dorantes-Aranda [1], John P. Bowman [2] and Gustaaf Hallegraeff [1]

[1] Institute for Marine and Antarctic Studies, University of Tasmania, 7004 Hobart, Australia; juan.dorantesaranda@utas.edu.au (J.J.D.-A.); Hallegraeff@utas.edu.au (G.H.)
[2] Tasmanian Institute of Agriculture, University of Tasmania, 7005 Hobart, Australia; john.bowman@utas.edu.au
* Correspondence: Aiko.Hayashi@utas.edu.au

Received: 8 November 2018; Accepted: 1 December 2018; Published: 8 December 2018

Abstract: Mycotoxins are emerging toxins in the marine environment, which can co-occur with algal toxins to exert synergistic or antagonistic effects for human seafood consumption. The current study assesses the cytotoxicity of the algal toxin okadaic acid, shellfish, and dust storm-associated mycotoxins alone or in combination on human intestinal (HT-29) and neuroblastoma (SH-SY5Y) cell lines. Based on calculated IC_{50} (inhibitory concentration 50%) values, mycotoxins and the algal toxin on their own exhibited increased cytotoxicity in the order of sydowinin A < sydowinin B << patulin < alamethicin < sydowinol << gliotoxin ≈ okadaic acid against the HT-29 cell line, and sydowinin B < sydowinin A << alamethicin ≈ sydowinol < patulin, << gliotoxin < okadaic acid against the SH-SY5Y cell line. Combinations of okadaic acid–sydowinin A, –alamethicin, –patulin, and –gliotoxin exhibited antagonistic effects at low-moderate cytotoxicity, but became synergistic at high cytotoxicity, while okadaic acid–sydowinol displayed an antagonistic relationship against HT-29 cells. Furthermore, only okadaic acid–sydowinin A showed synergism, while okadaic acid–sydowinol, –alamethicin, –patulin, and –gliotoxin combinations demonstrated antagonism against SH-SY5Y. While diarrhetic shellfish poisoning (DSP) from okadaic acid and analogues in many parts of the world is considered to be a comparatively minor seafood toxin syndrome, our human cell model studies suggest that synergisms with certain mycotoxins may aggravate human health impacts, depending on the concentrations. These findings highlight the issues of the shortcomings of current regulatory approaches, which do not regulate for mycotoxins in shellfish and treat seafood toxins as if they occur as single toxins.

Keywords: okadaic acid; sydowinin A; sydowinol; alamethicin; patulin; gliotoxin; combination index; synergy

Key Contribution: The present study, using human intestinal (HT-29) and neuroblastoma (SH-SY5Y) cell line models, demonstrated that okadaic acid–shellfish and –dust storm-associated mycotoxin combinations can result in synergistic toxic effects.

---

1. Introduction

The importance of fungi in the marine environment has been increasingly recognised in recent years. They are capable of infecting a wide range of marine animals, including sea turtles [1] and sea fan corals [2], and threatening human health through mycotoxin accumulation in seafood [3]. The majority of infectious fungi in the marine environment are considered to be of terrestrial origin [4], but atmospheric dust deposition and terrestrial runoff can facilitate the growth of fungi already residing in the marine environment and/or introduce them from terrestrial into marine environments. For example, an *Aspergillus sydowii* "bloom" (150,000 spores/m^2) along the east coast of Australia was observed after an extensive dust storm in 2009 [5]. Similarly, increased dust deposition and nutrient input from terrestrial runoff is thought to have contributed to an outbreak of the fungal disease sea fan coral aspergillosis in the Caribbean [6].

Fungal contaminants in seafood can also pose a significant human health risk. Several studies have shown that toxigenic fungal species can reside within the shellfish itself, seawater, and sediments from aquaculture farming areas. *Penicillium*, *Aspergillus*, *Trichoderma*, and *Cladosporium* have been isolated from such samples in France [7], Canada [8], Algeria [9], Russia [10], Brazil [11], Italy [12], and Tunisia [13]. These genera of fungi are capable of producing toxic metabolites (mycotoxins), including aflatoxins (AF), zearalenone (ZEA), deoxynivalenol (DON), fumonisins (FB), and ochratoxins (OTA) [14]. These compounds exhibit a wide range of biological activities, including hepatocarcinogenic, genotoxic, carcinogenic, oestrogenic, nephrotoxic, and nephrocarcinogenic effects [15]. Evidence exists that some shellfish-associated fungal isolates were capable of producing highly toxic mycotoxins, such as gliotoxin by *Aspergillus fumigatus* [3], patulin by *Penicillium* sp. [16], peptaibol by *Trichoderma* sp. [17], and griseofulvin by *P. waksmanii* [18]. These mycotoxins have been demonstrated to bio-accumulate in shellfish under both laboratory and natural conditions. A filtrate of marine-derived *T. koningii*, gliotoxin accumulated in shellfish, and peptaibols were detected in shellfish and sediments from aquaculture environments [3,17,19]. C17-sphinganine analogue mycotoxin (C17-SAMT) was claimed to be solely responsible for high shellfish toxicity in Tunisia in 2006 [13]. Mycotoxins are now widely viewed as new emerging toxins in shellfish.

Mycotoxins on their own can pose a significant health risk for humans through shellfish consumption, but an even greater concern arises from their possible synergistic effects with co-occurring algal toxins. However, mycotoxins in shellfish are currently not monitored and information on the combined effects of algal toxins and mycotoxins is sparse. So far, an in vivo Diptera larval bioassay by Ruiz et al. has been the only study to assess the combined effects of the algal toxin domoic acid and mycotoxin. Their study revealed increased toxicity by up to 34.5 times (the synergism factor) when domoic acid and longibranchi-A-I were injected together into Diptera larvae [20]. The proposed mechanism of this synergism was enhanced by an increase in Ca^{2+} influx into the cells by both domoic acid and novel peptaibol longibranchi-A-I [20].

The management of seafood safety is important for public health, market access, and public confidence. For example, a single incident of failure of detecting unacceptable levels of paralytic shellfish toxins (PST) in exported mussels resulted in AUD$24 million dollar economic loss to the Tasmanian seafood industry [21]. Current approaches to seafood safety management do not regulate for mycotoxins, and take no account of combined effects of co-occurring seafood toxins and treat them as if they were to occur as individual compounds [22,23]. Therefore, the aim of this study was to identify the toxic interactions of major algal toxins (e.g., saxitoxin, domoic acid and okadaic acid) and shellfish-associated (e.g., gliotoxins, patulin and peptaibol) and dust-originated (*A. sydowii* metabolites and sterigmatocystin [24]) mycotoxins (Figure 1) using human intestinal HT-29 and neuroblastoma SH-SY5Y cell line models. HT-29 and SH-SY5Y were chosen for assessing gastrointestinal and neurological effects, respectively, from saxitoxin [25], domoic acid [26] and okadaic acid [27]. Toxin interactions such as synergisms, antagonism, and additive were quantitatively evaluated with the combination index (CI) method [28].

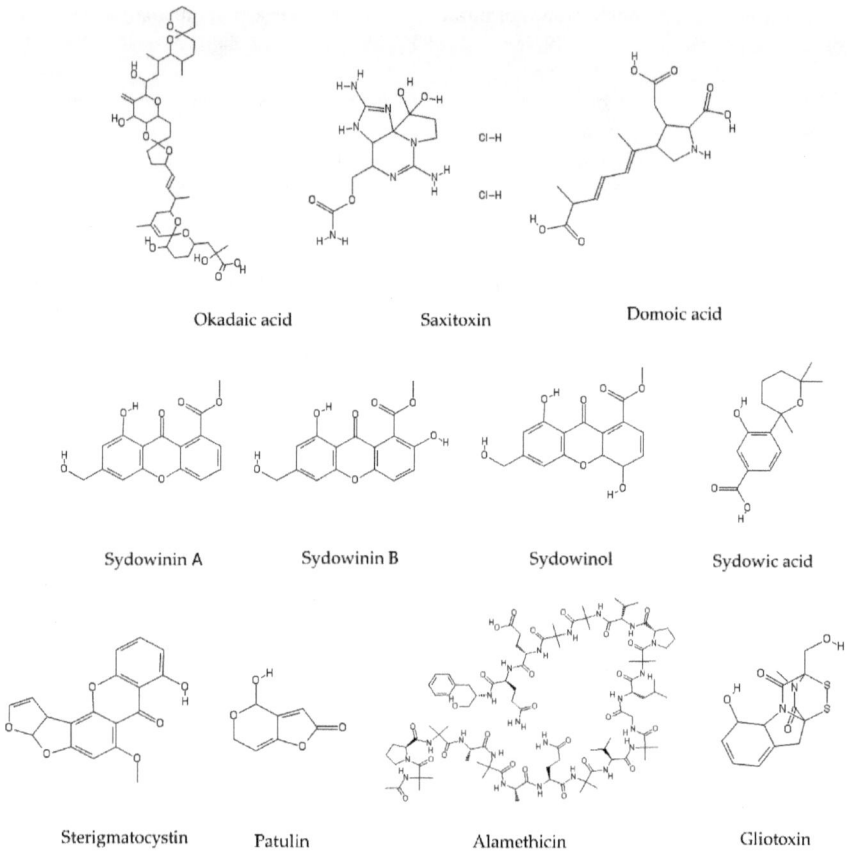

Figure 1. Chemical structures of typical algal toxins (okadaic acid, saxitoxin, domoic acid), dust storm-related mycotoxins (major *A. sydowii* metabolites and sterigmatocystin), and shellfish-related mycotoxins (patulin, alamethicin, gliotoxin).

2. Results

2.1. Individual Cytotoxicity of Algal Toxin and Mycotoxin

The cytotoxicity of individual mycotoxins and phycotoxins on the human intestinal cell line HT-29 and human neuroblastoma cell line SH-SY5Y was evaluated using resazurine cell viability reagent. The tested mycotoxins, except sydowic acid, exhibited a dose-dependent effect with a range of inhibitory concentration 50% (IC_{50}) from 65 nM to 124 µM for HT-29, and from 45 nM to 144 µM for SH-SY5Y (Table 1 and Supplementary Data in Figures S1 and S2). The tested concentration ranges of sydowic acid (HT-29: 0.028–283.75 µM, SH-SY5Y: 0.567–567.49 µM) showed no significant effect on viability for both HT-29 and SH-SY5Y (HT-29: $F_{(8,27)} = 0.095$, $p = 0.999$, SH-SY5Y: $F_{(4,15)} = 1.516$, $p = 0.248$). Sterigmatocystin reduced the viability of both HT-29 and SH-SY5Y in a dose-dependent manner with an incomplete sigmoid curve. The highest applicable concentration of 62 and 123 µM sterigmatocystin lowered the viability of HT-29 to 60%, and that of SH-SY5Y to 43%, respectively. Therefore, the IC_{50} of sterigmatocystin was not calculated. Okadaic acid displayed a dose-dependent effect on HT-29, with IC_{50} of 65 nM, and SH-SY5Y viability, with IC_{50} of 27 nM, whereas the other tested algal toxin, saxitoxin, and domoic acid had either no effect or minor effects on the viability of HT-29 and SH-SY5Y (maximum tested concentrations were 16.6–1.33 µM) (Figures S1 and S2). For the overall cytotoxicity ranking, based on the calculated IC_{50} values, the tested mycotoxin and algal toxin were found to be in the increasing order of sydowinin A < sydowinin B << patulin < alamethicin < sydowinol << gliotoxin ≈ okadaic acid in HT-29, and sydowinin B < sydowinin A << alamethicin ≈ sydowinol < patulin, <<gliotoxin < okadaic acid in SH-SY5Y.

Table 1. Summary of cytotoxicity of typical *A. sydowii* metabolites, dust storm/shellfish-associated mycotoxins, okadaic acid algal toxins on HT-29 and SH-SY5Y cells after 24 h exposure. Inhibitory concentration 50% (IC_{50}) values and 95% confidence interval (CI) were calculated from four replicates using the four-parameter logistic model (4PL) model.

Toxin	HT-29 IC_{50} (μM)	95% CI	SH-SY5Y IC_{50} (μM)	95% CI
Typical A. sydowii metabolites				
Sydowinin A	124.30	113.60–136.00	117.80	105.60–131.40
Sydowinin B	93.06	82.20–105.40	143.8	116.00–178.20
Sydowinol	2.50	2.21–2.82	5.14	5.06–5.23
Sydowic acid	NE (283.75) [1]	-	NE (283.75) [1]	-
Dust storm/shellfish mycotoxins				
Sterigmatocystin	>61.67 [2]	-	~123.35 [2]	-
Patulin	17.46	10.79–28.28	2.23	2.15–2.32
Alamethicin	4.92	4.57–5.29	5.43	5.29–5.67
Gliotoxin	0.062	0.052–0.075	0.045	0.039–0.053
Algal toxins				
Okadaic acid	0.065	0.056–0.075	0.027	0.026–0.029

[1] NE indicates toxins had no significant effect within the tested concentration range. Numbers in brackets indicate the maximum applicable concentration tested. [2] The maximum applicable concentration of 61.67 μM and 123.35 μM sterigmatocystin lowered the viability to 60% on HT-29 and 43% on SH-SY5Y, respectively.

2.2. Combined Cytotoxicity of Mycotoxins and Algal Toxin

Since okadaic acid was the only algal toxin which exhibited cytotoxicity on both HT-29 and SH-SY5Y cells, the effects of combined okadaic acid and mycotoxin sydowinin A, sydowinol, patulin, alamethicin, and gliotoxin on cell viability of HT-29 and SH-SY5Y were examined. Sydowinin B, sydowic acid, and sterigmatocystin were eliminated from the combined cytotoxicity assay because of their low cytotoxicity and limited solubility. Furthermore, the combination ratios were chosen to have an equipotent toxicity of each toxin (e.g., $(IC_{50})_1/(IC_{50})_2$ ratio) (Table 2), as there were no data available on the concentration of mycotoxins in shellfish, and this was recommended by Chou for an early stage study [29]. The combination index (CI) values were calculated from a fraction of cell viability affected (*fa*) values of 0.05 (corresponding to IC_{05}) to 0.97 (corresponding to IC_{97}), and the dose reduction index (DRI) was calculated when synergistic interactions were detected. All the binary mixtures of toxins showed a dose-dependent effect on HT-29 and SH-SY5Y cells (Figures S3 and S4).

Table 2. Molar combination ratio of okadaic acid and mycotoxin mixtures used in the assay.

Toxin Mixture	Molar Combination Ratio	
	HT-29	SH-SY5Y
Okadaic acid:Sydowinin A	1:1925.0	1:14366.2
Okadaic acid:Sydowinol	1:38.7	1:190.7
Okadaic acid:Alamethicin	1:76.6	1:201.1
Okadaic acid:Patulin	1:270.4	1:82.6
Okadaic acid:Gliotoxin	1:1.04	1:1.68

2.3. Okadaic Acid and Mycotoxins on Human Intestinal HT-29 Cells

Okadaic acid–sydowinin A, –alamethicin, –patulin, and –gliotoxin binary mixtures displayed variations of the interaction types on human intestinal HT-29 cells dependent upon the effect levels

(Figure 2). At low to moderate effect levels (*fa* < 0.65), these combinations exhibited antagonistic to additive effects, while they presented synergistic relationships at higher effect levels (*fa* > 0.65). In contrast to these okadaic acid–mycotoxin mixtures, okadaic acid–sydowinol mixtures displayed antagonistic effects at *fa* > 0.95 and a nearly additive interaction at *fa* < 0.95 (Figure 2). The DRI values for okadaic acid and mycotoxins varied from 1.8 to 12.5 and 1.8 to 12.2, respectively (Table 3). The greatest synergistic effect at *fa* = 0.9 was noted for the binary mixture of okadaic acid and gliotoxin, with a CI value of 0.41. For this combination, at the effect level of 0.9, the okadaic acid and gliotoxin mixture was 12.4 times more potent than okadaic acid alone, and 3 times more effective than gliotoxin alone.

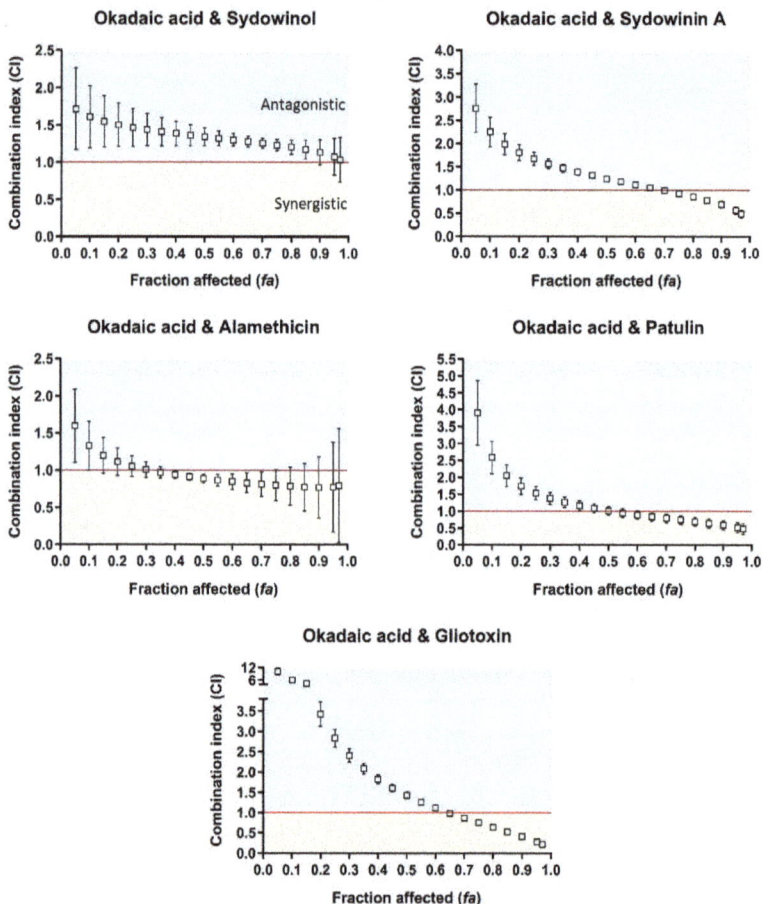

Figure 2. Combination index (CI)–fraction affected (*fa*, indicating fraction of cell viability affected. *fa* = 0.05–0.97 corresponds to 5–97% toxicity) curves for binary mixtures of okadaic acid and sydowinol, sydowinin A, alamethicin, patulin, and gliotoxin against human intestinal HT-29 cells. CI < 1, CI = 1, and CI > 1 indicate synergistic (orange rectangle), additive (red line), and antagonistic (blue rectangle) effects of binary mixtures, respectively. The error bar indicates 95% confidence intervals calculated using sequential deletion analysis (SDA).

Table 3. Combination index (CI) and dose reduction index (DRI) values for okadaic acid and mycotoxin combinations in HT-29 and SH-SY5Y cells at various effect levels (IC_{25}, IC_{50}, IC_{75} and IC_{90}). DRI values were only calculated when synergistic effects were detected. DRI implies fold of dose reduction for a given effect in a combination of toxins compared with the dose of each toxin alone.

Toxin Mixture	CI at				DRI at			
	IC_{25}	IC_{50}	IC_{75}	IC_{90}	IC_{25}	IC_{50}	IC_{75}	IC_{90}
HT-29								
Okadaic acid	1.67	1.24	0.93	0.69	-	-	2.72	3.61
Sydowinin A					-	-	1.80	2.43
Okadaic acid	1.47	1.34	1.23	1.13	-	-	-	-
Sydowinol					-	-	-	-
Okadaic acid	1.06	0.88	0.78	0.72	-	2.41	2.18	1.98
Alamethicin					-	2.14	3.14	4.61
Okadaic acid	1.53	1.01	0.75	0.53	-	-	1.76	1.95
Patulin					-	-	5.63	12.23
Okadaic acid	2.84	1.42	0.75	0.41	-	-	4.80	12.45
Gliotoxin					-	-	1.85	3.02
SH-SY5Y								
Okadaic acid	0.72	0.69	0.67	0.65	2.98	3.09	3.20	3.32
Sydowinin A					2.65	2.72	2.79	2.86
Okadaic acid	1.34	1.33	1.34	1.34	-	-	-	-
Sydowinol					-	-	-	-
Okadaic acid	1.30	1.33	1.36	1.41	-	-	-	-
Alamethicin					-	-	-	-
Okadaic acid	1.29	1.30	1.32	1.34	-	-	-	-
Patulin					-	-	-	-
Okadaic acid	1.30	1.48	1.68	1.91	-	-	-	-
Gliotoxin					-	-	-	-

2.4. Okadaic Acid and Mycotoxins on Human Neuroblastoma SH-SY5Y Cells

Okadaic acid–sydowinol, –alamethicin, –patulin, and –gliotoxin mixtures on human neuroblastoma SH-SY5Y cells showed an antagonistic interaction type at all effect levels, except that at *fa* = 0.05; gliotoxin and okadaic acid exhibited an additive interaction type (Figure 3). The calculated CI values for these combinations varied from 1.15 to 2.21 (Figure 3). By contrast, okadaic acid–sydowinin A mixtures exhibited synergisms at all effect levels, with a CI of 0.65 at *fa* = 0.9. For this combination, at the effect level of 0.9, the okadaic acid and sydowinin A mixture was 3.3 times more effective than okadaic acid alone and 2.9 times more effective than sydowinin A alone (Table 3).

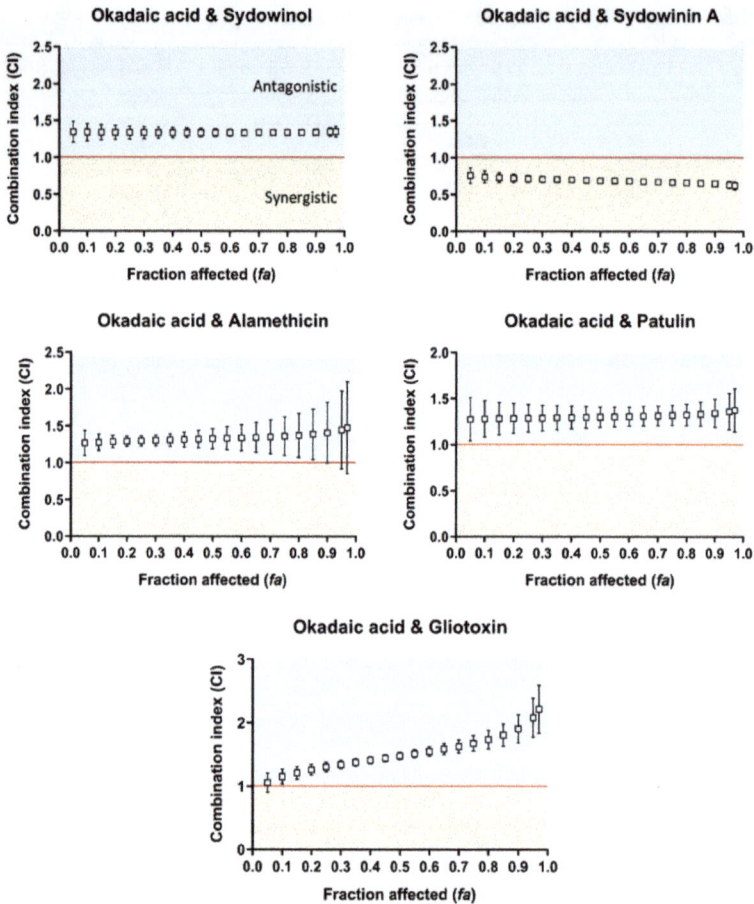

Figure 3. Combination index (CI)–fraction affected (*fa*, indicating fraction of cell viability affected. *fa* = 0.05–0.97 corresponds to 5–97% toxicity) curves for binary mixtures of okadaic acid and sydowinol, sydowinin A, alamethicin, patulin, and gliotoxin against human neuroblastoma SH-SY5Y cells. CI < 1, CI = 1, and CI > 1 indicate synergistic (orange rectangle), additive (red line), and antagonistic (blue rectangle) effects of binary mixtures, respectively. The error bar indicates 95% confidence intervals calculated using sequential deletion analysis (SDA).

3. Discussion

We demonstrated in this study that binary mixtures of the phycotoxin okadaic acid, and dust- and shellfish-associated mycotoxins exhibited cell line- and concentration-dependent antagonistic or synergistic interactions. Combinations of okadaic acid–sydowinin A, –alamethicin, –patulin, and –gliotoxin exhibited synergisms at higher effect levels and antagonisms at lower effect levels on HT-29. Interestingly, only okadaic acid–sydowinin A displayed synergism, whereas antagonism was noted for other combinations on SH-SY5Y at all effect levels. DRI values indicated that toxin doses can be theoretically reduced by up to 1.8 to 12-fold for the combination to have the same effect as that induced by each toxin on its own. These findings suggested that ingestion of a regulatory safe level of the algal toxin okadaic acid (0.16 mg OA eq./kg) could result in a health impact due to synergism with mycotoxin.

3.1. Synergisms between Okadaic Acid and Mycotoxins

We speculate that synergistic effects of okadaic acid and the tested mycotoxins on HT-29 could be the result of the impairment of cell structure. Okadaic acid is the main lipophilic marine biotoxin produced by *Dinophysis* and *Prorocentrum* dinoflagellates and responsible for diarrhetic shellfish poisoning (DSP) in humans [30]. Okadaic acid is an inhibitor of serine/threonine protein phosphatases (PP), which affect various important cellular metabolic processes, leading to cytoskeleton and intestinal mucosa deterioration, digestive dysfunction, lipid metabolism disorders, oxidative stress, and cellular apoptosis [31]. These series of events contribute to the gut barrier impairment and intestinal cell degeneration, which results in human diarrheic symptoms [31]. The mycotoxin alamethicin, also known as peptaibol, forms pores in membranes, thereby increasing membrane permeability [32]. Similarly, gliotoxin specifically binds to cytoplasmic membrane thiol groups, causing an increase in membrane permeability by affecting membrane protein orientation [33]. Patulin also induces the depletion of nonprotein sulfhydryl groups and increases potassium efflux, which results in the loss of structural integrity of the plasma membrane [34]. While mycotoxins have different mechanisms of action, they all lead to a disruption of ion homeostasis and structural damage which in turn potentially compounds downstream effects caused by okadaic acid in particular cytoskeleton deterioration, oxidative stress, and apoptosis. Furthermore, the observed shifts from antagonism to additive/synergism with increasing concentrations in the current study have also been reported in the similar study, where the interaction types of lipophilic phycotoxins (e.g., okadaic acid, pectenetoxin-2, yessotoxin, spirolide-1) were examined [35].

Okadaic acid and sydowinin A exhibited synergistic effects on both the HT-29 and SH-SY5Y cell lines. Currently, we lack knowledge of the details of the mode of action of the major *Aspergillus sydowii* metabolites sydowinin A and sydowinol. Sydowinin A has been reported to have more potent immunosuppressive effects on the Con A-induced and lipopolysaccharide-induced proliferations of mouse splenic lymphocytes compared to other *A. sydowii* metabolites [36]. The current study and other studies supported evidence of that the okadaic acid-induced PP inhibition also induces various neurotoxic effects [37,38]. However, no major human neurotoxic symptoms from ingesting okadaic acid-contaminated seafood have been reported so far, probably due to the levels of okadaic acid accumulating more slowly in the brain compared to the stomach and gastrointestinal tract tissues [39]. Synergistic relationships between okadaic acid and sydowinin A may have a basis in that the immunosuppressive characteristics of sydowinin A could sensitise cells to okadaic acid, but this requires investigation. The observed synergistic relationships with sydowinin A imply that even a low level of okadaic acid may cause significant neurotoxic effects in humans.

3.2. Antagonisms between Okadaic Acid and Mycotoxin on SH-SY5Y

The combination of okadaic acid and the tested mycotoxins exhibited antagonistic relationships against SH-SY5Y neuroblastoma cells, whereas interactions were synergistic against HT-29 intestinal cell lines. Antagonistic interactions were also noted for HT-29 at the low effect level. These observed antagonisms could be explained by multidrug resistance (MDR). MDR is regulated by P-glycoprotein (P-gp), which functions as an efflux transport pump, removing toxins from the plasma membrane, hence reducing cytotoxicity [40]. Okadaic acid efflux occurred in okadaic acid-resistant Chinese hamster ovary cells with increased levels of P-gp [41]. Therefore, the observed antagonisms in SH-SY5Y cells could be related to less mycotoxin binding to the target site, while okadaic acid is actively removed from the plasma membrane. This could lead to lower toxicity than estimated for the combined effect. This is supported by the fact that undifferentiated SH-SY5Y cells expressed some degree of P-gp expression, while HT-29 showed no detectable P-gp [42,43]. Furthermore, in the present study, mycotoxins were more abundant than okadaic acid in the binary mixtures, which could make mycotoxins more readily bind to the target site. Similarly, Alassane-Kpembi et al. (2015) suggested that the MDR drug efflux mechanism might explain the observed antagonism between deoxynivalenol

(DON)–3-Acetyldeoxynivalenol (3-ADON) and DON–Fusarenon-X (FX) combinations [44]. However, the suggested mechanisms of antagonisms remain speculative and require further study.

4. Conclusions

The present study demonstrated that binary mixtures of okadaic acid and shellfish- and dust-associated mycotoxins displayed cell line- and concentration-dependent interactions. The general interaction patterns observed in this study were a shift from antagonism to synergism with increasing concentrations on HT-29 cells, and antagonism or synergism at all concentrations on SH-SY5Y cells. The synergistic effects observed in the current study are of practical significance. While diarrhetic shellfish poisoning from okadaic acid and analogues is widely considered to be a comparatively minor seafood toxin syndrome (e.g., no human fatalities have ever occurred), our human cell model studies provided preliminary insights that synergisms with mycotoxins can be expected to more seriously aggravate human health impacts.

This also suggests the need for implementing more studies of seafood where there is risk of the co-occurrence of mycotoxins and algal toxins. Our results clearly demonstrate that the toxin interaction type depends on the effect level and cell type. This points to difficulties of predicting toxin interactions from the known mechanisms of actions of individual toxins without actual experimental data [29]. Mycotoxins are emerging toxins in seafood, and their occurrence may increase due to increased terrestrial runoff, dust storms, and the use of mycotoxin contaminated aquaculture feeds [23]. The current study did not explore the precise cellular mechanisms behind the mycotoxin and algal toxin interaction, and suggested mechanisms therefore remain speculative, and deserve further study. Future work should prioritise determining the interaction types of commonly occurring algal toxins (e.g., saxitoxin and domoic acid), and other mycotoxins (e.g., DON, AF, ZEA, FB, and OTA) [45]. Multiple mixtures (e.g., more than two toxins) should also be considered. Our results highlight the possible risks of toxin co-occurrence in seafood, a scenario which is not considered in current shellfish safety regulations.

5. Materials and Methods

5.1. Cell Line Cultures

Human neuroblastoma SH-SY5Y was kindly provided by Ms Yilan Zhen and Dr. Kaylene Young (Menzies Institute for Medical Research, University of Tasmania, Australia). Human colorectal adenocarcinoma cells HT-29 were kindly provided by Dr. Anthony Baker (Tasmanian Institute of Agriculture, University of Tasmania and School of Land and Food, Australia). Both cell lines were routinely maintained in Dulbecco's Modified Eagle's Medium (DMEM, D0819, Sigma-Aldrich, Sydney, Australia) supplemented with 10% foetal bovine serum (FBS, Bovogen Biologicals, Melbourne, Australia), and 100 U/mL penicillin and 100 mg/mL streptomycin solution in a humidified incubator (5% CO_2, 37 °C). SH-SY5Y cells were routinely subcultured at a ratio of 1:30–1:50, and medium changeover occurred approximately every 5 d. HT-29 cells were routinely subcultured at a ratio of 1:3–1:8, and medium changeover occurred approximately every 4 d.

5.2. Mycotoxin and Phycotoxin Toxins

Four typical *Aspergillus sydowii* metabolite standards, sydowinin A, sydowinin B, sydowinol, and sydowic acid were kindly provided by Professor Hiromitsu Nakajima, Tottori University, Japan. These compounds were isolated from *A. sydowii* IFO 4284 and IFO 7531 strains. Full descriptions of UV, IR, and NMR spectra, chemical structures, and molecular weights of these metabolites were previously provided by Hamasaki et al. (1975a,b) [46,47]. The crystallised *A. sydowii* metabolites were weighted on a microbalance and dissolved in small volumes of acetone (>0.5 mL). Among the other fungal toxins tested, gliotoxin (G9893, Sigma-Aldrich) was dissolved in ethanol, alamethicin (A4665, Sigma-Aldrich) was dissolved in DMSO, and sterigmatocystin (S3255, Sigma-Aldrich) and patulin

(P1639, Sigma-Aldrich) were dissolved in acetonitrile. Phycotoxin standards, saxitoxin (CRM-STX-f), domoic acid (CRM-DA-g), and okadaic acid (CRM-OA-d) were purchased from the National Research Council Canada. Concentrations used are expressed as µM.

5.3. Cytotoxicity Bioassays

When cells reached >70% confluency, they were detached using a trypsin–EDTA solution. Detached cells were centrifuged 300 g for 5 min and resuspended. Cells were seeded to a 96-well plate at 1.0×10^4 cells/well for HT-29 and 3.0×10^4 cells/well for SH-SY5Y and allowed to attach for 24 h prior to toxin exposure. Each well contained 100 µL of cells suspension, and 0.5–3% (v/v) of algal toxin and mycotoxins stocks were added to the basal DMEM, which contained no supplemented FBS nor antibiotics. Concentration ranges of tested individual toxicity of algal toxins and mycotoxins were 1.33×10^{-9}–123.3 µM for SH-SY5Y, and 3.12×10^{-8}–235.6 µM for HT-29. For the combined cytotoxicity bioassay, the ranges were 0.019–214.9 µM for HT-29 and 0.016–169.6 µM for SH-SY5Y. Cells were rinsed once with DPBS (Dulbecco's phosphate-buffered saline, 0.9 mM $CaCl_2$; 0.50 mM $MgCl_2 \cdot 6H_2O$; 2.7 mM KCl; 1.5 mM KH_2PO_4; 137.9 mM NaCl; 8.1 mM $Na_2HPO_4 \cdot 7H_2O$). Toxin-containing DMEM was added to each well and incubated further for 24 h. Controls received only solvents, and the solvent concentration used in the assay was preliminary tested to have no significant effect on the cell viability compared to those received basal DMEM without solvents (data not shown). After toxin exposure, the cells were washed once again with DPBS and 100 µL of the same basal media (without phenol red) containing 5% resazurin solution [48] were added to each well. Following additional 2 h incubation in the dark, the plate was read with a BMG FLUOstar OMEGA plate reader using excitation of 540 nm and emission of 590 nm. Cell viability was expressed as the percentage of fluorescence reading compared to the control (% of control). Four replicates were prepared for each treatment.

5.4. Statistical Analysis of Cytotoxicity of Individual Mycotoxin and Algal Toxin

Data analysis was conducted with the decision tree proposed by Sérandour et al. [49], except that in this experiment, the controls were preliminary tested to have no effect on cell viability and no further calculation was conducted when there was no bottom asymptote. Briefly, the dose response curves were fitted with the four-parameter logistic model (4PL), and 95% asymptotic confidence intervals were calculated using GraphPad Prism 7. The half-maximal inhibitory concentration (IC_{50}) indicating the concentration that caused a half-maximal viability was calculated for each toxin. IC_{50} was accepted if the fitting dose–response curve had $R^2 > 0.85$ and standard of error of log IC_{50} was <40%. One-way analysis of variance (ANOVA) was used to evaluate statistical differences between control and treatments. Tukey's honestly significant different (HSD) post hoc tests were performed when the main effect was significant. Appropriate data transformation was determined using Box–Cox transformation. ANOVA and follow-up statistical analyses were performed with the statistical software R (R Development Core Team, version 3.4.3) [50]. A significance level of 0.05 was applied in this study.

5.5. Median Effect and Combination Index Analysis of Mycotoxin and Algal Toxin Mixture

The cytotoxicity of mycotoxin and algal toxin mixture was analysed based on the Chou–Talalay method [28]. The combination of mycotoxin and algal toxin were at an equipotency ratio (e.g., $(IC_{50})_1/(IC_{50})_2$ ratio) based on the calculated IC_{50} values using the graphpad prism 4PL model; therefore, each toxin roughly affects the cell viability equally [29]. The dose–responses for individual toxins and their mixture were modelled using the median effect equation of the mass action law:

$$\frac{fa}{fu} = \left(\frac{D}{D_m}\right)^m \tag{1}$$

where D is the dose of the toxin, D_m is the median effect dose (e.g., IC_{50}), fa is the fraction affected by dose (D) (e.g., fractions of cell viability affected), fu represents the fraction unaffected, and m indicates the shape of the slope ($m = 1, > 1$, and < 1 indicate hyperbolic, sigmoidal, and flat sigmoidal curves, respectively). Toxin interactions were only analysed when the linear correlation coefficient (r) of the median effect plot was greater than 0.92.

The mycotoxin and algal toxins interaction was analysed by the combination index (CI) method derived from the median effect equation of the mass action law. The combination index was calculated using the following equation below [29]:

$$^n(CI)_x = \sum_{j=1}^{n} \frac{(D)_j}{(D_x)_j} \tag{2}$$

where $^n(CI)_x$ is the combination index for n mycotoxins and algal toxins that inhibits x percent of a system (e.g., viability), $(D)_j$ are the doses that mixture of n phyco- and mycotoxins that inhibits x percent of a system, and $(D_x)_j$ are the doses that each phyco- and mycotoxin itself inhibits x percent of a system. CI $< 1, = 1$, and > 1 indicate synergism, additive effect, or antagonism, respectively. CI values were calculated over a range of $fa = 0.05$ to 0.97 (5–97% toxicity). A confidence interval of 95% (95% CI) for CI was calculated based on sequential deletion analysis (SDA). The dose reduction index (DRI) values were determined for the combination that exhibited a synergistic relationship at IC_{25}, IC_{50}, IC_{75} and IC_{90}. DRI indicates the magnitude of how the dose of each drug in a mixture can be reduced at the given effect level compared to the doses of each drug alone. The dose–response analyses of toxin mixtures, CI, and DRI were performed with Compusyn software (ComboSyn Inc., Paramus, NJ, USA).

Supplementary Materials: The following are available online at http://www.mdpi.com/2072-6651/10/12/526/s1, Figure S1: Dose–response curves of (a) major *A. sydowii* metabolites, (b) dust storm/shellfish mycotoxins, and (c) algal toxins on human intestinal HT-29 cells. Data are mean ± SD of four replicates. Figure S2: Dose–response curves of (a) major *A. sydowii* metabolites, (b) dust storm/shellfish mycotoxins, and (c) algal toxins on human neuroblastoma SH-SY5Y cells. Data are mean ± SD of four replicates. Figure S3: Okadaic acid (OA), sydowinol (SYD), sydowinin A (SYDA), alamethicin (ALA), patulin (PAT), and gliotoxin (GLI) and their binary mixture dose–responses for cytotoxicity against the human intestinal HT-29 cell line. Concentrations in combinations were expressed as the sum of the concentrations of two toxins. Data are mean ± SD of four replicates. Figure S4: Okadaic acid (OA), sydowinol (SYD), sydowinin A (SYDA), alamethicin (ALA), patulin (PAT), and gliotoxin (GLI) and their binary mixture dose–responses for cytotoxicity against the human neuroblastoma SH-SY5Y cell line. Concentrations in combinations were expressed as the sum of the concentrations of two toxins. Data are mean ± SD of four replicates.

Author Contributions: A.H. and G.H. developed the design and ideas of this work and wrote the manuscript with input from all authors. A.H. conducted experiments and analyzed the data. J.J.D.-A. helped to set up some preliminary experiments. J.P.B. provided experimental support and facilities.

Funding: This work was partially funded by Australia Research Council grant DP130102725.

Acknowledgments: Hiromitsu Nakajia, Tottori University, Japan, provided us with four typical *A. sydowii* metabolites. Yilan Zhen and Kaylene Young, Menzies Institute for Medical Research, University of Tasmania, Australia, provided us with human neuroblastoma SH-SY5Y cells. Anthony Baker, Tasmanian Institute of Agriculture, University of Tasmania and School of Land and Food, Australia, provided us with Human colorectal adenocarcinoma cells HT-29.

Conflicts of Interest: The authors declare no conflict of interest.

References

1. Sarmiento-Ramírez, J.M.; Abella, E.; Martín, M.P.; Tellería, M.T.; López-Jurado, L.F.; Marco, A.; Diéguez-Uribeondo, J. *Fusarium solani* is responsible for mass mortalities in nests of loggerhead sea turtle, Caretta caretta, in Boavista, Cape Verde. *FEMS Microbiol. Lett.* **2010**, *312*, 192–200. [CrossRef] [PubMed]
2. Smith, G.W.; Ives, L.D.; Nagelkerken, I.A.; Richie, K.B. Caribbean sea-fan mortalities. *Nature* **1996**, *383*, 487. [CrossRef]
3. Grovel, O.; Pouchus, Y.F.; Verbist, J.F. Accumulation of gliotoxin, a cytotoxic mycotoxin from *Aspergillus fumigatus*, in blue mussel (*Mytilus edulis*). *Toxicon* **2003**, *42*, 297–300. [CrossRef]

4. Pang, K.L.; Overy, D.P.; Jones, E.B.G.; da Luz Calado, M.; Burgaud, G.; Walker, A.K.; Johnson, J.A.; Kerr, R.G.; Cha, H.J.; Bills, G.F. 'Marine fungi' and 'marine-derived fungi' in natural product chemistry research: Toward a new consensual definition. *Fungal Biol. Rev.* **2016**, 1–13. [CrossRef]
5. Hallegraeff, G.; Coman, F.; Davies, C.; Hayashi, A.; McLeod, D.; Slotwinski, A.; Whittock, L.; Richardson, A.J. Australian Dust Storm Associated with Extensive *Aspergillus sydowii* Fungal "Bloom" in Coastal Waters. *Appl. Environ. Microbiol.* **2014**, *80*, 3315–3320. [CrossRef] [PubMed]
6. Harvell, C.D.; Kim, K.; Burkholder, J.M.; Colwell, R.R.; Epstein, P.R.; Grimes, D.J.; Hofmann, E.E.; Lipp, E.K.; Osterhaus, A.D.; Overstreet, R.M.; et al. Emerging marine diseases—Climate links and anthropogenic factors. *Science* **1999**, *285*, 1505–1510. [CrossRef]
7. Sallenave-Namont, C.; Pouchus, Y.F.; Robiou Du Pont, T.; Lassus, P.; Verbist, J.F. Toxigenic saprophytic fungi in marine shellfish farming areas. *Mycopathologia* **2000**, *149*, 21–25. [CrossRef]
8. Brewer, D.; Greenwell, M.; Taylor, A. Studies of *trichoderma* isolates from *mytilus edulis* collected on the shores of Cape Breton and Prince Edward islands. *Proc. N. S. Inst. Sci.* **1993**, *40*, 29–40.
9. Matallah-Boutiba, A.; Ruiz, N.; Sallenave-Namont, C.; Grovel, O.; Amiard, J.C.C.; Pouchus, Y.F.; Boutiba, Z. Screening for toxigenic marine-derived fungi in Algerian mussels and their immediate environment. *Aquaculture* **2012**, *342–343*, 75–79. [CrossRef]
10. Zvereva, L.V.; Vysotskaya, M.A. Filamentous fungi associated with bivalve mollusks from polluted biotopes of Ussuriiskii Bay, Sea of Japan. *Russ. J. Mar. Biol.* **2005**, *31*, 382–385. [CrossRef]
11. Santos, A.; Hauser-Davis, R.A.; Santos, M.J.S.; De Simone, S.G. Potentially toxic filamentous fungi associated to the economically important *Nodipecten nodosus* (Linnaeus, 1758) scallop farmed in southeastern Rio de Janeiro, Brazil. *Mar. Pollut. Bull.* **2017**, *115*, 75–79. [CrossRef]
12. Greco, G.; Cecchi, G.; Di Piazza, S.; Cutroneo, L.; Capello, M.; Zotti, M. Fungal characterisation of a contaminated marine environment: The case of the Port of Genoa (North-Western Italy). *Webbia* **2018**, *7792*, 1–10. [CrossRef]
13. Marrouchi, R.; Benoit, E.; Le Caer, J.P.; Belayouni, N.; Belghith, H.; Molgó, J.; Kharrat, R. Toxic C17-Sphinganine Analogue Mycotoxin, Contaminating Tunisian Mussels, Causes Flaccid Paralysis in Rodents. *Mar. Drugs* **2013**, *11*, 4724–4740. [CrossRef] [PubMed]
14. Gonçalves, R.A.; Naehrer, K.; Santos, G.A. Occurrence of mycotoxins in commercial aquafeeds in Asia and Europe: A real risk to aquaculture? *Rev. Aquac.* **2016**, 1–18. [CrossRef]
15. Zain, M.E. Impact of mycotoxins on humans and animals. *J. Saudi Chem. Soc.* **2011**, *15*, 129–144. [CrossRef]
16. Vansteelandt, M.; Kerzaon, I.; Blanchet, E.; Fossi Tankoua, O.; Robiou Du Pont, T.; Joubert, Y.; Monteau, F.; Le Bizec, B.; Frisvad, J.C.; Pouchus, Y.F.; et al. Patulin and secondary metabolite production by marine-derived *Penicillium* strains. *Fungal Biol.* **2012**, *116*, 954–961. [CrossRef]
17. Poirier, L.; Montagu, M.; Landreau, A.; Mohamed-Benkada, M.; Grovel, O.; Sallenave-Namont, C.; Biard, J.F.; Amiard-Triquet, C.; Amiard, J.C.; Pouchus, Y.F. Peptaibols: Stable Markers of Fungal Development in the Marine Environment. *Chem. Biodivers.* **2007**, *4*, 1116–1128. [CrossRef]
18. Petit, K.E.; Mondeguer, F.; Roquebert, M.F.; Biard, J.F.F.; Pouchus, Y.F. Detection of griseofulvin in a marine strain of *Penicillium waksmanii* by ion trap mass spectrometry. *J. Microbiol. Methods* **2004**, *58*, 59–65. [CrossRef]
19. Sallenave, C.; Pouchus, Y.F.; Bardouil, M.; Lassus, P.; Roquebert, M.F.; Verbist, J.F. Bioaccumulation of mycotoxins by shellfish: Contamination of mussels by metabolites of a *Trichoderma koningii* strain isolated in the marine environment. *Toxicon* **1999**, *37*, 77–83. [CrossRef]
20. Ruiz, N.; Petit, K.; Vansteelandt, M.; Kerzaon, I.; Baudet, J.; Amzil, Z.; Biard, J.F.; Grovel, O.; Pouchus, Y.F. Enhancement of domoic acid neurotoxicity on Diptera larvae bioassay by marine fungal metabolites. *Toxicon* **2010**, *55*, 805–810. [CrossRef]
21. Campbell, A.; Hudson, D.; McLeod, C.; Nicholls, C.; Pointon, A. Tactical Research Fund: Review of the 2012 paralytic shellfish toxin event in Tasmania associated with the dinoflagellate alga, *Alexandrium tamarense*. In *FRDC Project 2012/060 Appendix to the Final Report*; SafeFish: Adelaide, Australia, 2013.
22. Stobo, L.A.; Lacaze, J.P.C.L.; Scott, A.C.; Petrie, J.; Turrell, E.A. Surveillance of algal toxins in shellfish from Scottish waters. *Toxicon* **2008**, *51*, 635–648. [CrossRef]
23. Gonçalves, R.A.; Schatzmayr, D.; Hofstetter, U.; Santos, G.A. Occurrence of mycotoxins in aquaculture: Preliminary overview of Asian and European plant ingredients and finished feeds. *World Mycotoxin J.* **2017**, *10*, 183–194. [CrossRef]

24. Hayashi, A.; Crombie, A.; Lacey, E.; Richardson, A.; Vuong, D.; Piggott, A.; Hallegraeff, G. *Aspergillus Sydowii* Marine Fungal Bloom in Australian Coastal Waters, Its Metabolites and Potential Impact on *Symbiodinium* Dinoflagellates. *Mar. Drugs* **2016**, *14*, 59. [CrossRef]

25. Gessner, B.D.; Middaugh, J.P. Paralytic shellfish poisoning in alaska: A 20-year retrospective analysis. *Am. J. Epidemiol.* **1995**, *141*, 766–770. [CrossRef]

26. Teitelbaum, J.S.; Zatorre, R.J.; Carpenter, S.; Gendron, D.; Evans, A.C.; Gjedde, A.; Cashman, N.R. Neurologic Sequelae of Domoic Acid Intoxication Due to the Ingestion of Contaminated Mussels. *N. Engl. J. Med.* **1990**. [CrossRef]

27. Valdiglesias, V.; Prego-Faraldo, M.V.; Pásaro, E.; Méndez, J.; Laffon, B. Okadaic Acid: More than a diarrheic toxin. *Mar. Drugs* **2013**, *11*, 4328–4349. [CrossRef]

28. Chou, T.C.; Talalay, P. Quantitative analysis of dose-effect relationships: The combined effects of multiple drugs or enzyme inhibitors. *Adv. Enzyme Regul.* **1984**, *22*, 27–55. [CrossRef]

29. Chou, T.C. Theoretical Basis, Experimental Design, and Computerized Simulation of Synergism and Antagonism in Drug Combination Studies. *Pharmacol. Rev.* **2006**, *58*, 621–681. [CrossRef]

30. Food and Agriculture Organization (FAO). Marine biotoxins. In *FAO Food and Nutrition*; Food and Agriculture Organization of the United Nations: Rome, Italy, 2004; p. 80.

31. Wang, J.; Wang, Y.Y.; Lin, L.; Gao, Y.; Hong, H.S.; Wang, D.Z. Quantitative proteomic analysis of okadaic acid treated mouse small intestines reveals differentially expressed proteins involved in diarrhetic shellfish poisoning. *J. Proteomics* **2012**, *75*, 2038–2052. [CrossRef]

32. Mueller, P.; Rudin, D.O. Action potentials induced in biomolecular lipid membranes. *Nature* **1968**, *217*, 713–719. [CrossRef]

33. Jones, R.W.; Hancock, J.G. Mechanism of Gliotoxin Action and Factors Mediating Gliotoxin Sensitivity. *J. Gen. Microbiol.* **1988**, *134*, 2067–2075. [CrossRef]

34. Riley, R.T.; Showker, J.L. The Mechanism of Patulin's Cytotoxicity and the Antioxidant Activity of Indole Tetramic Acids. *Toxicol. Appl. Pharmacol.* **1991**, *126*, 108–126. [CrossRef]

35. Fessard, V.; Alarcan, J.; Barbé, S.; Kopp, B.; Hessel-Pras, S.; Lampen, A.; Le Hégarat, L. In vitro assessment of binary mixtures effects of phycotoxins in human intestinal cells. In Proceedings of the 18th International Conference on Harmful Algae, Nantes, France, 21–26 October 2018; p. 468.

36. Liu, H.; Chen, S.; Liu, W.; Liu, Y.; Huang, X.; She, Z. Polyketides with immunosuppressive activities from mangrove endophytic fungus *Penicillium* sp. ZJ-SY2. *Mar. Drugs* **2016**, *14*, 217. [CrossRef]

37. Tapia, R.; Peña, F.; Arias, C. Neurotoxic and synaptic effects of okadaic acid, an inhibitor of protein phosphatases. *Neurochem. Res.* **1999**, *24*, 1423–1430. [CrossRef]

38. Arias, C.; Sharma, N.; Davies, P.; Shafit-Zagardo, B. Okadaic acid induces early changes in microtubule-associated protein 2 and tau phosphorylation prior to neurodegeneration in cultured cortical neurons. *J. Neurochem.* **1993**, *61*, 673–682. [CrossRef]

39. Matias, W.G.; Traore, A.; Creppy, E.E. Variations in the distribution of okadaic acid in organs and biological fluids of mice related to diarrhoeic syndrome. *Hum. Exp. Toxicol.* **1999**, *18*, 345–350. [CrossRef]

40. Lum, B.L.; Gosland, M.P. MDR expression in normal tissues. Pharmacologic implications for the clinical use of P-glycoprotein inhibitors. *Hematol. Oncol. Clin. N. Am.* **1995**, *9*, 319–336. [CrossRef]

41. Tohda, H.; Yasui, A.; Yasumoto, T.; Nakayasu, M.; Shima, H.; Nagao, M.; Sugimura, T. Chinese hamster ovary cells resistant to okadaic acid express a multidrug resistant phenotype. *Biochem. Biophys. Res. Commun.* **1994**, *203*, 1210–1216. [CrossRef]

42. Breuer, W.; Slotki, I.N.; Ausiello, D.A.; Cabantchik, I.Z. Induction of multidrug resistance downregulates the expression of CFTR in colon epithelial cells. *Am. J. Physiol.* **1993**, *265*, C1711–C1715. [CrossRef]

43. Bates, S.E.; Shieh, C.Y.; Tsokos, M. Expression of mdr-1/P-glycoprotein in human neuroblastoma. *Am. J. Pathol.* **1991**, *139*, 305–315.

44. Alassane-Kpembi, I.; Puel, O.; Oswald, I.P. Toxicological interactions between the mycotoxins deoxynivalenol, nivalenol and their acetylated derivatives in intestinal epithelial cells. *Arch. Toxicol.* **2015**, *89*, 1337–1346. [CrossRef]

45. Chou, T.C. Drug combination studies and their synergy quantification using the chou-talalay method. *Cancer Res.* **2010**, *70*, 440–446. [CrossRef]

46. Hamasaki, T.; Sato, Y.; Hatsuda, Y. Structure of Sydowinin A, Sydowinin B, and Sydowinol, Metabolites from *Aspergillus sydowi*. *Agric. Biol. Chem.* **1975**, *39*, 2341–2345. [CrossRef]

47. Hamasaki, T.; Sato, Y.; Hatsuda, Y. Isolation of new metabolites from *Aspergillus sydowi* and structure of sydowic acid. *Agric. Biol. Chem.* **1975**, *39*, 2337–2340. [CrossRef]
48. O'Brien, J.; Wilson, I.; Orton, T.; Pognan, F. Investigation of the Alamar Blue (resazurin) fluorescent dye for the assessment of mammalian cell cytotoxicity. *Eur. J. Biochem.* **2000**, *267*, 5421–5426. [CrossRef]
49. Sérandour, A.L.; Ledreux, A.; Morin, B.; Derick, S.; Augier, E.; Lanceleur, R.; Hamlaoui, S.; Moukha, S.; Furger, C.; Biré, R.; et al. Collaborative study for the detection of toxic compounds in shellfish extracts using cell-based assays. Part I: Screening strategy and pre-validation study with lipophilic marine toxins. *Anal. Bioanal. Chem.* **2012**, *403*, 1983–1993. [CrossRef]
50. R Core Team. *R: A Language and Environment for Statistical Computing*; R Foundation for Statistical Computing: Vienna, Austria, 2017.

MDPI

St. Alban-Anlage 66

4052 Basel

Switzerland

Tel. +41 61 683 77 34

Fax +41 61 302 89 18

www.mdpi.com

Toxins Editorial Office

E-mail: toxins@mdpi.com

www.mdpi.com/journal/toxins

www.ingramcontent.com/pod-product-compliance
Lightning Source LLC
Chambersburg PA
CBHW051914210326
41597CB00033B/6144